Do I Have an Environmental Disease?

Walter Wortberg

Do I Have an Environmental Disease?

**Recognition and Prevention of the Causes
of Cancer and Chronic Diseases**

Case Studies and Investigations from a General and Environmental
Physician's Practice, with New Diagnostic and Therapeutic Approaches

Bibliographic Information published by the Deutsche Nationalbibliothek
The Deutsche Nationalbibliothek lists this publication in the Deutsche Nationalbibliografie; detailed bibliographic data is available in the internet at http://dnb.d-nb.de.

Library of Congress Cataloging-in-Publication Data
Wortberg, Walter, 1938- , author.
 [Bin ich umweltkrank? English]
 Do I have an environmental disease? : recognition and prevention of the causes of cancer and chronic diseases : case studies and investigations from a general and environmental physician's practice, with new diagnostic and therapeutic approaches / Walter Wortberg.
   p. ; cm.
 Translation of: Bin ich umweltkrank? : Die Ursachen von Krebs und chronischen Krankheiten erkennen und vermeiden / Dr. Walter Wortberg.
 Includes bibliographical references.
 ISBN 978-3-631-66247-2 (print) — ISBN 978-3-653-05703-4 (ebook)
 I. Title. [DNLM: 1. Environmental Health—Case Reports. 2. Chronic Disease—prevention & control—Case Reports. 3. Environmental Medicine—Case Reports. 4. Environmental Pollution—Case Reports. WA 30.5]
 RB152.5
 616.9'8—dc23   2015011537

Publication of the English translation in agreement with Mediengruppe
Oberfranken – Fachverlage GmbH & Co. KG, Kulmbach

Cover Image:
© coldwaterman - Fotolia.com_47990248

ISBN 978-3-631-66247-2 (Print)
E-ISBN 978-3-653-05703-4 (E-Book)
DOI 10.3726/978-3-653-05703-4

© for the English Edition: Peter Lang GmbH
Internationaler Verlag der Wissenschaften
Frankfurt am Main 2015
PL Academic Research is an Imprint of Peter Lang GmbH.
© for all other Languages: ML Verlag in der Mediengruppe Oberfranken –
Fachverlage GmbH & Co. KG, Kulmbach

Peter Lang – Frankfurt am Main · Bern · Bruxelles · New York ·
Oxford · Warszawa · Wien

All parts of this publication are protected by copyright. Any utilisation outside the strict limits of the copyright law, without the permission of the publisher, is forbidden and liable to prosecution. This applies in particular to reproductions, translations, microfilming, and storage and processing in electronic retrieval systems.

This publication has been peer reviewed.
www.peterlang.com

Dedication

My beloved family
My grandchildren Noah, Eliah, Liuba, and Milian
and my children Daniel and Hee-Jeung, Hendrik, Nikola, and Burkard,
who often had to make do without their grandfather and father;
my wife Monika, who was my tireless proofreader and time and again
put me back on the narrow road towards the realistically doable
and who helped me to formulate my message
so as not to step on anybody's toes.

Contents

1.	A greeting to a more human world (Dr. P. Jennrich) ...	13
2.	A plea for more fairness in environmental justice (Prof. Dr. E. Schöndorf)	15
3.	Prologue	19
4.	Introduction: why this book?	21
5.	Why does environmental medicine get so little attention in Germany?	29
6.	Consequences of our exploitation of nature	35
6.1.	Poisonings of plants and animals	35
6.2.	Human poisoning due to industrial accidents	37
6.3.	Health risks due to nuclear accidents and temporary storage ...	39
6.4.	Other harmful substances with creeping health effects	41
	6.4.1. Drinking water and nutrition	41
	6.4.2. Energy-saving lamps	42
	6.4.3. Contaminated soil, contaminated water, poisoned waste disposal sites	42
6.5.	Electromagnetic radiation, mobile communications, electrosmog	46
	6.5.1. Recommendations: what can we do?	53
6.6.	The connection between infectious diseases and metals	54
6.7.	Gene technology: curse or rescue?	54
6.8.	Conventional medicine at a crossroads—side-effects	56
	6.8.1. Medical drugs	56
	6.8.2. Medical products	60
	6.8.3. Vaccinations	65
6.9.	Animal testing as pointless exploitation of nature ...	67
7.	Case studies from my own practice	69
7.1.	Introduction: from my own life	70
7.2.	Case studies	72

7.2.1.	The first spark: poisoning by dental restoration material	73
7.2.2.	Poisoning caused by intrauterine damage to the foetus	77
7.2.3.	An example of expert witness testimony	92
7.2.4.	Depression and asthma: reimbursing costs of medically required dental restoration work is denied	103
7.2.5.	Huntington's disease	105
7.2.6.	Malignant brain tumor (glioblastoma)	108
7.2.7.	Acute allergic reaction to metals (hypersensitivity) after surgery	110
7.2.8.	Six breast operations due to mammary carcinoma	111

8. Eight environmental-medical studies from daily practice ... 117

8.1. Evaluating a questionnaire from KV Dortmund ... 117
    8.1.1. Introduction ... 117
    8.1.2. Method ... 118
    8.1.3. Results ... 118
    8.1.4. Critical assessment ... 121

8.2. Dangers posed by dental restoration material ... 123
    8.2.1. Introduction ... 123
    8.2.2. Method ... 123
    8.2.3. Results ... 124
    8.2.4. Critical assessment ... 130

8.3. Intrauterine foetal damage caused by maternal exposure to heavy metals ... 134
    8.3.1. Introduction ... 134
    8.3.2. Method ... 135
    8.3.3. Results ... 136
    8.3.4. Critical assessment ... 138

8.4. Health damage from tin compounds ... 140
    8.4.1. Introduction ... 140
    8.4.2. Method ... 142
    8.4.3. Results ... 142
    8.4.4. Critical assessment ... 143

8.5. Environmental diseases in 360 patients with heavy metal exposure ... 145
    8.5.1. Introduction ... 145

	8.5.2.	Method	146
	8.5.3.	Results	146
8.6.	Heavy metal and harmful substance exposure in six cases of rare diseases of unknown etiology		148
	8.6.1.	Progressive muscle dystrophy	148
	8.6.2.	Psoriatic arthritis	148
	8.6.3.	Tinnitus	149
	8.6.4.	Miscarriages and stillbirths	149
	8.6.5.	Endometriosis	150
	8.6.6.	Neurodermitis	150
	8.6.7.	Critical assessment	151
8.7.	Exposure and hypersensitivity to metals in 139 patients with a benign tumor		152
	8.7.1.	Method	152
	8.7.2.	Results	152
8.8.	Influence of heavy metals and immunologic and genetic factors on the development of malignant tumors		152
	8.8.1.	Introduction	152
	8.8.2.	Results	155
	8.8.3.	Critical assessment	157
8.9.	Twelve years later: recognition of brain damage from environmental toxins		159
	8.9.1.	Introduction: why am I presenting this case?	159
	8.9.2.	Some remarks to the patient's prior history	160
	8.9.3.	Findings	161
	8.9.4.	The patient's application for rehab-measures	163
	8.9.5.	Critical Assessment	164
	8.9.6.	My reasons for publicizing this case	168
	8.9.7.	Summary of the results of the studies	168
9.	**Environmental-medical diagnostics**		**171**
9.1.	Introduction: patients and their own interest in environmental medicine		171
9.2.	Internal exposure to metals		172
	9.2.1.	Nonmetallic materials (plastics, ceramics, endoprostheses)	175
9.3.	External exposure to metals		176
	9.3.1.	Nutrition, drinking water, soil, and air	176
	9.3.2.	Clothing, toys, coins	176

9.4.	Interior and exterior exposure to other industrial products and fungi toxins	177
9.5.	The importance of a patient's prior history and of clinical examinations	178
9.6.	The difference between acute and chronic poisoning	180
9.7.	Symptoms and chronic diseases due to exposure to harmful substances	180
	9.7.1. Examination for external exposure (environmental monitoring)	184
	9.7.2. Examination for internal exposure (biomonitoring)	185
	9.7.3. Testing for genetic disorders	196
	9.7.4. Imaging as diagnostic method	197
	9.7.5. Exposure to harmful chemicals and fungi toxin	197
9.8.	Alternative medicine—drugs and treatments without side effects?	198
	9.8.1. Dental organogram	198
9.9.	Experience and its implications for diagnostics	202

10. Therapeutic options for treating environmental diseases ... 205

10.1.	Introduction: so you are suffering from an environmental disease. What now?	205
10.2.	Detoxification with the help of drugs	207
	10.2.1. Detoxification after heavy metal exposure	207
	10.2.2. Detoxification for industrial exposure	210
	10.2.3. Dental restoration and detoxification	211
	10.2.4. Apheresis: only in the most severe cases	213
	10.2.5. Treatment of fungal toxin poisoning	215
10.3.	Detoxification with side-effect-free medication—naturopathy	216
	10.3.1. Healing stones	217
	10.3.2. Black Serpent Stones	220
	10.3.3. Microalgae	221
	10.3.4. Homeopathy	223
	10.3.5. Acupuncture	224
10.4.	Environmental medicine is holistic medicine	226

11. A healthy lifestyle: what can we do ourselves? ... 227

11.1.	Introduction: what does "healthy" mean?	227
11.2.	Nutrition	228

11.3.	Sports, exercise, gymnastics	229
11.4.	Massage	230
11.5.	Psyche—the significance of our souls	231
11.6.	Meditation	233
11.7.	Hobbies	234
11.8.	Bio-meditation—bioenergy: a higher-level therapy	235

12. Course of action: the seven steps ... 237

13. Critical assessment and conclusions ... 241

13.1.	Introduction	241
13.2.	Medical assessment	241
13.3.	Legal assessment	252
13.4.	Moral and ethical assessment	255

14. Excerpt from an open letter by Prof. Wassermann ... 261

15. My wishlist for the future ... 263

15.1.	A greeting from the "Natur & Heilen" (90.) publishing company	264
15.2.	An appeal, and a "Thank you!"	264

16. A few words for you to take home with you (Prof. Dr. med. Frentzel-Beyme) ... 267

17. Appendix ... 269

17.1.	Some words of thanks	269
17.2.	Explanations and definitions	270
17.3.	Addresses: associations, laboratory institutes, clinics	273
17.4.	Bibliography	277

1. A greeting to a more human world (Dr. P. Jennrich)

We hold in our hands the fruit of a committed general and environmental physician's 25 years of experience. Now, everyone who is prepared to widen their horizon and to question deadlocked dogmas can take part in this experience.

Walter Wortberg's book is an urgent plea for more justice and humanity in medicine, something from which each single individual can profit just as much as our healthcare system as a whole.

In addition, Walter Wortberg demonstrates that as a civilization, we are almost ready to saw off the tree branch on which we sit—we are, however, no longer using a simple hand saw but, following the spirit of our time, attack it with a chainsaw. Everybody can do his part to help effect the necessary and long-overdue global changes. This with-and-for-each-other starts with how we treat nature, animals, and our fellow human beings.

I wish for this book, which grants us insights into its author's life's work, to effect those changes that are so close to the author's heart.

Peter Jennrich
Director of the International Board of Clinical Metal Toxicology
Scientific Advisor to the German Medical Board for Clinical Metal Toxicology

2. A plea for more fairness in environmental justice (Prof. Dr. E. Schöndorf)

In his first chapter, the author of this text puts forth his primary question as to why environmental medicine is given so little attention in Germany. As former public prosecutor, I would like to look at this question from a legal point of view, exaggerated a little for polemical purposes. Why are none of those who, through their pollution of the environment or their marketing of toxic products, physically abuse people or damage their health, in any of our prisons (§ 223 StGB—German Penal Code)? As a rule, environmental diseases are related to harmful substances—they don't come out of the blue but are caused by human interaction with the environment which, in turn, may attract the attention of criminal law.

In the following, I would like to answer my question with the help of an example. This case, which concerned me during my time as environmental public prosecutor in the 1980s and 1990s, provides some interesting insights into the environmental-medical attention deficit disorder—rightfully decried by the author—that extend beyond the strictly legal dimension. The case in question is the so called Frankfurt wood preservative trial. For 13 years, I was in charge of the investigations into this case. I apologize in advance that, in the context of this preface, I will have to reduce the comprehensive investigations to only a few points. I do, however, believe that those points can help in appreciating this book.

In 1984, a patient's initiative, representing a few dozens of families from all over Germany, brought charges of bodily injury due to the use of toxic wood preservatives. The plaintiffs claimed that people became seriously ill after treating the wood panelling in their houses with these questionable chemicals. After five years of investigation I became convinced that I found proof of biocidal ingredients of the preservatives—most importantly the fungal toxin PCP—causing illness in a large number of people. Surprisingly, however, my charges against both managing directors of the manufacturer were rejected. The argumentation: if toxicologists are arguing, the legal system must show restraint and should not try to be the referee.

Regardless of the question as to whether the legal system should indeed be as humble as that, this point reveals environmental law and environmental health's Achilles heel, namely in causality. There is no single environmental controversy without two diametrically opposed parties. No matter if we are dealing with amalgam or softening agents, mobile communications, silicon implants, or the present case of wood preservatives, they are subjects of highly controversial debate. This fact is quite understandable, at least from a legal point of view. Where classic bodily injury crimes involve fists, baseball bats, or knives, the environmental field has to deal with a different situation, where the 'weapons' are not palpable, nor can we taste or smell them, and they are invisible, because they act on the nanoscale. In human blood, we could only measure a few millionth of a gram of the ultra-toxic dioxin contained in the wood preservative. In the light of this highly complicated problem of causality, with its often hard-to-prove correlations, environmental medicine necessarily has a tough job and must fight for acceptance.

In 1992 and 1993, after I successfully appealed the rejection of the charges described above, the main trial took place. The testimony of a Swiss toxicology, selected by the defence and the reputed PCP-pope, was heard and considered the *non plus ultra* with regard to the wood preservative industry sector. His testimony made clear that the aforementioned problem of causality was not limited to the context of the natural sciences but also has a political side to it—after all, the Swiss toxicologist issued certificates stating that the wood preservatives are absolutely harmless to human health. His utter belief was based on an experiment with rat food, which the animals survived despite high doses of PCP in it. That which does not damage rats will not damage humans—it's as simple as that, according to the man from Zürich. What his report did not mention, however, was that where the rats received the poison through their food, inhabitants of the affected houses had ingested the active ingredients through their respiratory system, which from a toxicological point of view is much more dangerous. In addition, ultra-toxic dioxin had been removed from the PCP before feeding it to the rats. This was never done to any of the agents on sale in home stores. And finally, all classic psychiatric/neurological complaints, such as lack of drive, concentration lapses, and loss of strength, as expressed by the residents, were never in any way considered in the animal study. The study concentrated only on the inner organs, which showed—for whatever reason—no significant changes.

There was no doubt that the professor from Switzerland had lied. Whenever an issue concerns mass market products, which earn millions for their manufacturers, toxicologic expert witness involves the work of con men, who are commissioned by the manufacturers and paid large amounts of money to certify these products. The consequences are dramatic: due to the lack of sufficiently substantiated dangers to public health, politicians and the legal system see no reason for acting and consumers too, due to the maze of conflicting information, have no grounds for changing their behaviour. These all are further reasons as to why the "toxic party" carries on undisturbed. Nevertheless, the Frankfurt Environmental Court did not believe the PCP-pope and in June 1993 convicted the defendant of causing bodily harm.

Finally: the appeal. The defense rebuked that during the trial statements were given by a biased expert. The physician from Heidelberg had testified that these wood preservatives must be considered highly dangerous to anybody who is exposed to them. After the charges were initially rejected, out of "professional concern", he had encouraged me to continue trying to bring the case to trial; by that time he had "[...] unambiguously diagnosed more than 70 patients with health damage due to wood preservatives". The defense presented this letter as proof of the bias of this expert. The German Federal Court of Justice (BGH) took the same view and overruled the verdict made by the Frankfurt Environmental Court. Germany's highest judges had turned professional concern and medical responsibility into arguments for prejudice of an expert witness. For many critics this reasoning was just a cheap pretext for repealing the verdict that more than 250,000 victims had been waiting for for such a long time.

Why does the Federal Court of Justice bend the law? I allow myself to provide an answer. The crux of the situation is to hide and deny the dark sides of our system—a system that is built on technical progress and growth and consciously accepts all the risks that come with it. An honest inventory of the damage we did to our environment would quickly reveal that this is not about negligible collateral damage, but rather concerns fundamental construction faults in our system. This error is not only morally unjustifiable but is also financially unsustainable and therefore puts the whole system at risk. The wood preservatives case discussed above, only one of many environmental controversies, cost a quarter of a million people their health and caused 365 billions (in German Marks) of damages. Had the facts that were found been determined in a legally binding manner, it would have created a problem for the chemical establishment—and maybe not only for them. We are—still—afraid of these kinds of consequences. Politicians and the legal system still do their utmost to obscure causal connections, so that the harsh truth will never see the light of day. With this in mind, it is understandable that both environmental justice and environmental medicine have large obstacles to overcome before they can truly take off. However, the sign of the times is that things will change and that the lock will spring once critical mass is reached. I am positive that this book, written by such a committed environmental physician as Walter Wortberg, will be of lasting help.

Prof. Dr. Erich Schöndorf
Former public prosecutor and prosecutor in the Frankfurt wood preservative trial.

3. Prologue

My book
it tells the history of the suffering of millions of people with environmental health damage,
it provides hope to millions of people with environmental health damage,
it looks over its shoulder back into history.

It is a shout to the heavens that must shake up everybody,
it is hunger for more justice,
it is a longing for more humanity.

It analyzes the present,
it connects ancient wisdom and modern science,
it connects humankind and nature.

It is a challenge that must be realized,
it enlightens the layman, so in need of information,
it appeals to all responsible political and industrial parties.

The message: take care!

> *"To reach the source, one must go against the flow."*
> Confucius

4. Introduction: why this book?

*"What we know is a drop,
what we don't know is an ocean."*
Isaac Newton

There are three reasons that led me to writing this book:

1. Do I have an environmental disease?
Am I suffering from an environmental disease? During the past 25 years, more and more patients traveling from all over Germany asked me this question when visiting my practice for general, environmental, and tropical medicine. The question was put into their mind through the internet, neighbours, television, their family physician, and my lectures. An overwhelming number of people asked me for help after hearing one of my lectures. Although I have been retired since 2003, my telephone does not cease to ring. This public interest in environmental medicine stands in shrill contrast with the official reaction from the authorities. I have been supervising self-help groups, unions, and networks consisting of victims of environmental health damage. What do they suffer from? To answer this question, I have to digress a little further and take you, the reader, off to Africa.

During my time in a bush hospital in Akwanga, Nigeria, from 1971 to 1974, I never saw many of the diseases that we encounter in the industrialized countries—P. Jennrich (58.) calls them "diseases of affluence", with which he wants to express that they are a consequence of our contemporary lifestyle, which in turn is an emergent property of our industrialization: it shapes us more strongly than we want to see, both in positive and in negative ways. Because these diseases concern everybody, they are diseases of affluence. How does one recognize these "diseases of affluence"?

There are no typical signs or specific complaints. These diseases are often characterized by highly stubborn symptoms that improve little, despite intensive treatment from conventional medical practitioners. Maybe this is even a good thing and a clever idea by "mother nature". Symptoms of pain, itching, reddening and flaking of the skin, vertigo, tingling sensations in the legs, and fatigue are warning signs. They protect us against larger dangers to our health and signal that something is not alright with our body.

Exposure to harmful substances must be considered as a cause for any pain or suffering that continues over months or even years—i.e., chronic diseases. Once a connection can be shown, we know we are dealing with an environmental disease.

What are the most frequently encountered symptoms and diagnoses (i.e., diseases)? Some example:

Organ	Psychiatric	Nervous system	Inner organs/ immune system	Hormonal system
Symptoms	rapid fatigue	headaches	chronic coughing	disturbances of the menstrual cycle
	inner turmoil	problems concentrating	fever episodes of unknown etiology	fertility disorders
	insomnia	memory lapses	elevated liver enzymes	thyroid inflammations
	depressive disorder	feelings of numbness	relapsing infections	thyroid nodules
	anxiety	vertigo	elevated liver values (transaminases)	
	no longer able to work under pressure	tingling sensation in the legs		
		mental and physical impediments		

What many people don't want to know is that this group of diseases of affluence also includes:

- skeletal and connective tissue problems: chronic rheumatoid arthritis, fibromyalgia, osteoporosis, arthritis
- Cardiovascular diseases: high blood pressure, arteriosclerosis, cardiac diseases
- Pediatric diseases: headaches, migraine, AD(H)D, neurodermitis, eczema, autism
- Diseases that lead to mental and physical disability (130., 134., 134.)

- Finally, this group also includes cancer, the most devastating disease of all. In his book (84.) S. Mukherjee calls it "The King of Diseases". Cancer is frequently the final station in a long life of suffering.

We can expand this list at will, but all of these diseases share a common property: their prevalence has increased dramatically over the past 20 to 30 years: their frequency has doubled, sometimes even tripled and quadrupled. Growth rates as large as this cannot be explained away by increased life expectancy and increasing professional stress.

In 2005 and 2009 my wife and I visited Burkina Faso and Tanzania (in West and East Africa, respectively), where we were sad to find out that our diseases of affluence are also on the advance in these countries. Their causes are, without any doubt, chemicals and pesticides, i.e., many harmful substances that have been banned in Germany for years. Now, they are exported to the developing countries. What an outrage!

It hurts. Africa is, as it were, my second home country: both of our two children were born there.

2. Passing on our knowledge—more human care instead of technology and profit

For me who has occupied himself with the topic of environmental medicine for the past 40 years, in my capacity as private physician, scientist, and expert witness, it is almost my duty to pass on my knowledge to victims of environmental damage—for them, there is no lobby. With this book, I want to tell them that they are not left alone with their problems.

In my general and environmental health practice, between 1987 and 2003, I treated 11,600 patients, 1,600 of them (i.e., 15 %) suffering from clinically proven environmental damage. After this, because of my retirement, I have stopped counting. This was end of August 2003.

All patients underwent conventional as well as environmental-medical examinations and received appropriate treatment. My studies extended over many years. To begin with, I chronicled the design for my proceedings—in technical jargon this is called the study design. There exist few studies of this kind, which is regrettable since only with the help of this type of study we can arrive at persuasive results.

As early as in 1994, 1999, 2002, and 2003 I documented and sent the first results of my environmental-medical studies to the state and federal medical boards and the corresponding odontological boards. In these documents, I urgently pointed out the dangers posed by harmful substances. I also sent copies to the public health insurance companies and health ministries on state as well as federal level (a total number of ca. 50). Since 1998, I hold lectures for the medical boards and environmental committees of the association of public physicians Dortmund (KV Dortmund) and the Westfalen/Lippe medical association in Münster—something that my environmental-medical colleagues also do. All of us have shown great engagement with our environmentally damaged patients. However, all our efforts have so far not had any result.

In the summer of 2009 I personally handed one of these files with documentation to the president of the medical professional association, and three weeks later to the chairman of the KV Dortmund. I added a list of participants in an environmental health conference in Trier. Both of them gave me an almost identical reaction: "I have never seen any environmentally damaged patient. How does one recognize them?"

Only the State medical association of Westfalen/Lippe answered me with a written statement that proudly pointed out their Environmental Information Van, which could be called on demand. I received this answer with a shake of my head. It made both myself and many of the colleagues whom I mentioned this statement ask ourselves if our new environmental-medical findings are completely ignored. I cannot believe this nor do I want to!

In the meantime I have published part of my results in scientific journals. All patients gave their informed consent and reaching as many people as possible was a crucial factor in publishing my results in the form of this book—something that is not the case for publication in professional journals. Even though the populace has a marked interest in environmental medicine, there is a rather small level of concrete knowledge amongst them.

Our medical profession must once again become more human. Technology and logic by themselves are not sufficient for practicing medicine: it is an art of healing that today regrettably is almost lost, says B. Lown (74.) in his book "The Lost Art of Healing".

Healing as an art needs quietness, patience, and time.
It must be practiced with passion and love.

In the end, healing cannot only be explained by biochemistry and biophysics, even though our greatest contemporary scientists believe this. Healing means to reach the soul of a human being. With this, we enter the realm of metaphysics that cannot be grasped by the intellect alone. A human being consists of body/matter, mind/intellect, and soul. Today, we have scientific proof that psychological problems can lead to genetic disorders.

We must remove the underlying causes, the origin of these toxic effects. This way, we have to keep people in good health for as long as possible rather than to elaborate on ever more individual symptoms—something which in the end will no longer be financially sustainable. Informing the public is the foremost precondition for reaching this target.

During my time in Africa I founded a nursing school. The first ever lesson started out with the baffling question posed by a 14 year old boy as to "what is the mother of the disease?". This is the central question of the medical profession, namely what is the cause of disease. I continually had to think of that.

In his book, "The Secret of Healing. How Ancient Wisdom Changes Medicine", J. Faulstich (29.) showed that there is no contradiction between ancient wisdom and modern science. Reading this book effectively gave me the courage to dare undertake my own balancing act of writing a book that targets both laymen and conventional medical practitioners and to have the courage to describe ancient wisdom and modern medical science next to each other, as equal partners.

3. A close acquaintance with a severe environmental disease as the actual motivation for this book

My actual motivation for scientifically documenting and exactly evaluating all environmental-medical cases that I encountered was a visit to my practice, in 1987, by a close acquaintance, who presented a classic example of environmental damage due to inappropriate treatment with dental restoration material (chapter 7.2.1).

I will not try to hide that I am aiming for understanding and support from in particular my younger colleagues. Since devoting myself ever more to environmental medicine my life has gained lots of new stimuli. That which applies to the doctor also applies to the patient when regarding the disease as new life experience and as a sign of personal reorientation. The cases I discuss in chapter 7.2 provide signs of everything that is possible.

From a scientific point of view it is clear that my single study will not be able to ruttle the foundations of our classic, conventional medicine. Hence, I decided to sketch out my 25 years of experience with and research into environmental diseases. In this, I lay even higher worth on my own experiences than on the experimental results—in the end, experiences cannot be wrong, they are always true and hence truthful. Also, experiences gained in one's own life are easily understood by an individual layman. When presenting them in conjunction with results from objective research, one has passed on all information that one is humanly capable of.

In chapter 5 I investigate why environmental medicine gets so little attention in Germany.

In chapter 6 I describe some industrial accidents to illustrate the consequences of our exploitation of nature. The consequential damage to plants, animals, and also to humans make the necessity of re-appreciating environmental medicine as its own scientific field blatantly obvious.

Chapter 7 lists several case studies. Here, some of the people concerned might recognize themselves, although I obviously changed the names of the patients. With the help of these examples, I sketch out a first set of diagnostic and therapeutic options, which enable the layman to find, with the help of their physician, the best diagnostic and therapeutical methods. Only the attending physician, who knows their patient well, can select the correct approach and determine the right dosages—they should act the coordinator during the whole course of treatment, and in their hands all strains should converge.

In chapter 8 I present results from studies in my own practice in the form of individual papers that prove that these examples are not isolated cases. I obtained permission to conduct tests on hospital patients, supported by the respective specialist departments, from the Westfalen/Lippe medical association's ethical commission. Each article ends with a critical assessment, which may involve an intentional partial overlap and repetition of the same statements. Because the matter at hand will be unfamiliar to most readers, I opt to introduce it to them step by step.

In chapters 9 to 12 I will report new diagnostic and therapeutic treatment options. By taking note of these procedures, patients are able to better understand the cause of their illness and the therapeutical measures taken. This way, the patient is also capable of influencing their progress. A lot of space is occupied by the question as to how to remain fit without taking medical drugs. I will present concrete steps as to how to proceed.

Chapter 13 contains a critical assessment of all that was discussed before, together with the necessary conclusion. Here, I refer back to the findings from previous chapters in order to integrate them into the higher-level social and sociopolitical situation. It is not possible to practice effective environmental medicine without this high-level overview. With it in mind, I finally present my wish list.

I hope that this book is written in such a way that it is understandable to the layman. I abstained from presenting expansive numerical data in tables, without scientific rigor suffering from it. All tabulated material can be found in the individual publications or requested from me directly. Once the underlying cause has been found, difficult correlations become easy to understand. However, finding those causes often requires the skills of a private investigator.

When using technical jargon, I opted to use the German term followed by the jargon term in parenthesis, for example: poisons (noxa) and cancer (carcinoma). It is of regrettable importance for laymen to learn technical terms, since we as medical professionals prefer to use jargon even when talking to our patients.

This book has no intention to cause harm to anyone. The seriousness of its topic virtually commands honesty and fairness. Nevertheless, if one suffers together with patients, in particular victims of environmental disease, for several decades, one unwittingly assimilates their emotions, disappointments, the way they talk, and their reproaches, which can then be reflected in the text. On the other hand, I do not hide the fact that the decline of contemporary medicine worries me immensely. Only clear words and ruthless exposure of its weaknesses can stop this decline and show it the way out of the dead end street in which it has found itself.

> *"Today's sufferers of environmental diseases are only snowballs compared to tomorrow's avalanche."*
> *Conclusion of the Symposium held by the non-profit network for patients diagnosed with environmental disease (Gemeinnütziges Netzwerk für Umwelt-Kranke Genuk e. V.), 21.4.2012*

5. Why does environmental medicine get so little attention in Germany?

Take your health into your own hands!

To all victims of environmental diseases, one would like to shout out: "Take your health into your own hands!". This book sets out to show that this is both possible and necessary. This regrettable necessity is because, at this time, there is no support to be expected from politics, industry, and the public health insurance companies—a frustrating conclusion that I reached after numerous discussions with colleagues in private healthcare, senior physicians from clinics and university hospitals, and the responsible parties in our healthcare system. Their responses were of a single voice: "We no longer have the financial means or the time for scientific research into the causes of environmental diseases."

Fifty years ago hardly anyone mentioned health damage due to harmful substances. Although their consequential damage has seen a dramatic increase and their causes have in all cases been established by scientifically qualified research, they are given very little official attention.

Even the Environmental Party (The Greens) have forgotten about environmental medicine long ago. I vividly remember how, under Federal Chancellor Schröder's new government, two Green Party ministers, Andrea Fischer and Jürgen Trittin, took over the Ministry for Public Health (BMG) and the Ministry for Environmental Affairs (BMU), both very important with regard to public health. The environmentalist movements and NGOs (non-governmental organizations) were filled with euphoria and believed a rethinking in matters of exploiting the environment—keywords: nuclear energy/environmental medicine—would happen.

Once they took their seats on the governmental benches, the Greens once again lost sight of their voters' health. K. Müller (85.), a committed environmental physician and founder of the German environmental physician's professional association (Deutscher Berufsverband der Umweltmedizin—DBU) and the European Academy for Environmental Medicine, is correct when he writes in the headline to one of his articles that "in the public health policies of our parties, environmental medicine does not play any role."

Clearly the responsible parties suffer from a lack of awareness of the health damage caused by harmful substances from the environment.

What do we mean with the term environmental medicine? When observing the many definitions that we have, we rapidly see that we are dealing with a big complex of diseases and their causes, all of which have different definitions. The European Academy for Environmental Medicine and the Austrian Medical Association define environmental medicine as follows:

> *Environmental medicine is the science of prevention, diagnosis, and treatment of diseases correlated with environmental factors. One part of environmental medicine focuses on prevention where the other is a more clinical approach (26.)*

In the Bavarian physician's continued-education-policy from October 1, 1993 (German medical journal, 1996), environmental medicine is defined as follows: "Environmental medicine encompasses the medical care for single individuals who suffer from health problems or who underwent examinations with suspicious findings, all of which are associated with environmental factors by either themselves or their attending physician." (24.)

I consider the definition provided by the European Academy for Environmental Medicine the simplest one and hence the one most easily understood by the layman. In addition, the definition is quite broad, something which, as a holistic medical physician, is close to my heart.

However, since preventive and clinical environmental medicine are inseparably intertwined, I also regard them as a unit. Hence, I always treated them as equals both in my thinking and in my action.

The fact that environmental medicine hardly plays any significant role is not only valid for Germany, but also for other industrialized countries. In his article "20 Years of Environment and Health, from the Point of View of an NGO", E. Petersen (96.) provided a striking treatment of this subject. He presents the results from five European conferences on environment and health in almost minute detail. It starts with the first Conference in Frankfurt am Main (Germany) in May 1989, followed by conferences in Helsinki (Finland) in June 1994, London (UK) in June 1999, Budapest (Hungary) in June 2004, and in Parma (Italy) in March 2010.

The series of conferences is sponsored by the European division of the WHO (World Health Organization), which is, so to speak, the highest constitutive health authority. Hence, the preconditions for better dealings with our environment and hence better environmental health for each individual were perfectly satisfied. At their first conference in Frankfurt, in 1989, the European charter for environment and health expressed it thusly: "Each human being has the right to an environment conducive to the highest amount of health and wellbeing."

One cannot find a formulation with better implications for the individual citizen. Regrettably, though, the exact opposite happened: none of the health conferences managed to change anything. Here, we certainly must also blame the scientific establishment, which is researching the wrong part of the problem—instead of researching the causes, it concentrates on pathological-physiological processes and on symptom elimination. Despite all progress we made in understanding these processes, we should remain humble and say, with Newton, "What we know is a drop, what we don't know is an ocean."

In his book "Humanity's Final Storage Facility", W. Prada (98.) formulates it even more clearly when, in light of the increasing number of diseases and catastrophes, he writes: "Our error is that we think that we can control everything with our technology and logical thinking. The opposite is the case."

The examples provided by the devastating nuclear accidents in Chernobyl and Fukushima have not sufficiently irritated the responsible parties. Other than clever slogans, no real help is to be expected from that corner (see chapter 6).

With this in mind, each individual must search for the causes of his complaints by himself and ask: why am I sick? One must ask oneself: am I prepared to research the cause of my disease? Am I prepared to draw the necessary consequences with regard to my lifestyle, nutrition, etc.?

> "He who demands good health must be prepared to leave behind his sickness-inducing lifestyle."
> Franz Schmaus

Franz Schmaus is an international expert on healing and founder of Myko-Troph, an institute for nutritional science and mycology which resides in the Hessian city of Limeshain-Rommelhausen. He is an expert on the traditional Asian art of healing and testifies to the effectiveness of healing mushrooms.

> Considering the threat posed by environmental harmful substances, the question appears as to the Why of my disease. That which is valid for each individual and their disease also applies to the whole field of environmental medicine. To remedy anything, one must remove its cause.

Why does environmental medicine live the life of a wallflower? Why are the pioneers of the field so often laughed at and even treated as crackpots? In the following section, I would like to discuss the reasons for this phenomenon.

It is a fact that despite enormous technological progress during the past 20 years our healthcare system is going downhill—progress in medical technology regrettably is no longer affordable to many of us. This is felt particularly poignantly by the patients and family physicians at the base of the pyramid—progress at the expense of the "little man", one might say. Especially in these circumstances we must not lose track of environmental science, which so far covers only a small area of medical science. There is no money to research new territory.

We must regard all of medicine as part of our whole social and political context. The political landscape is determined by financial crises, energy crises, and public debts which, due to globalisation, affect every single country.

In his book "Means and Measures: the Fight Towards Appropriate Organization", H. Münkler (88.) establishes the pursuit of security as one of humankind's most basic needs. Hence, every European and western government puts military and economic security in the foreground. Our commercial industry advertises with new scientific and technological breakthroughs all the time. Apparently this solves all our problems—gene technology is the magic word. Freighter ships, cargo airliners, and luxury cruise liners become ever larger and faster. Hence, the individual's health and wellbeing necessarily suffer. The best example is provided by the USA, where there is still no public health insurance system, something which exists in, wait and listen, the smallest remote village in Siberia. Americans interpret mandatory health insurance as socialism. According to many of them, individual freedom includes the freedom to not insure against becoming ill but instead to carry the risk by oneself (76.).

In principle, we must consider an environmentally-related health disorder as the cause of any diseases of unknown etiology, i.e., diseases for which conventional medicine cannot find a cause. When we apply this principle,

80 % to 90 % of the time we'll find ourselves on the right track towards a diagnosis. It is exactly this track, this necessary lifeline, that is ignored by all responsible parties, not only in Germany but in every industrialized country.

We can only speculate about the reasons as to why in Germany so little attention is paid to environmental medicine (these reasons are, incidentally, also valid for all industrialized nations).

1. Environmental medicine is not commonly taught and researched in universities. Only a few academic or university hospitals have a department for environmental medicine. In Germany, the field is classified under public hygiene and formal training to qualify as an environmental physician takes five years. But which doctor attends an institute for public hygiene and then starts a private medical practice?! The consequence: there are not enough qualified medical physicians.
2. Environmental medicine means prevention. Prevention does not earn money.
3. Only in a rare few cases the costs of environmental-medical treatment are carried by the public health insurance companies.
4. If a patient suspects an environmental disease as cause of their occupational invalidity, they must provide 100 % proof of a causal connection between disease and environmental toxin—a proof that is both legally and medically accepted.
5. There is no interest from the pharmaceutical industry to remove the causes of disease. One may ask oneself where we would end up, economically, if everybody were healthy.
6. The situation is aggravated by the fast-paced developments made during the past 20 to 40 years by the chemical industry, who present us with countless new chemicals and new, supposedly improved drugs and medical products. Apparently, it is impossible to thoroughly test all these products for human tolerability.

In summary, we might say that if we believe our leading environmental scientists, the future of our beautiful planet Earth looks seriously threatening. "Even though our legislators tolerate solvents, exposure to heavy metals, exhaust fumes, diesel soot, ozone, insecticides, herbicides, fungicides, azodyes, softening agents, stabilizing agents, flavor enhancers and emulsifiers, food preservatives, antioxidants, synthetic fragrances, radioactive radiation, nitrate, electrosmog, and mobile communication technology. Our body, however, doesn't tolerate them." (51.)

In light of the dramatic increase in diseases of affluence, legislators must ask themselves if this "silent tolerance" can today still be justified politically and morally. For the most part, the keywords that rule our economy still are: growth, sales increase, rationalization.

> *More growth and boosting the economy are the favorite demands made by top-ranking representatives in the European Union. Rarely if ever does quality get mentioned.*

For thousands of years, humans lived in harmony with nature only to have become its exploiter. Industrialization caused radical changes in the relation between Man and his environment. Hence, we find ourselves at a crossroads.

Do we want to continue this reckless exploit and hence run the risk of poisoning ourselves, or do we want to treat our natural resources in such a way that we leave our offspring an environment worth living in? There is still time to decide against poisoning and our future should belong to prevention.

> *Environmental medicine is prevention.*

According to my medical and human understanding it is a physician's task and duty to point out these dangers. Once health damage becomes irreparable, there is not much that a medical doctor can do. This may sound exaggerated, but there is truth in it: we as doctors must care more about prevention, i.e., we must do everything within our power to prevent diseases from development. However, we lack the necessary conditions to achieve this—preventive medicine in the truest sense is unwanted and, individual cases excepted, is not supported.

The dramatic rise in diseases of affluence, weather catastrophes, and the unimpaired rise in emission of CO_2 and other harmful substances should give us pause for thought and urge us to act quickly. If we don't, the consequences of our destruction of nature will overrun our pursuit of military and economic security.

6. Consequences of our exploitation of nature

6.1. Poisonings of plants and animals

As medical professionals, we are only a small wheel in the giant mechanism of our environment. However, even the biggest engine can only run as long as its smallest component remains functional. Every one of us is responsible for the good shape of their own part of this machinery and must ensure that it runs in the face of adversity.

In this light, we may ask ourselves how we can reduce the risks posed to our health and the environment by in particular new products, in these at the moment so unfavourable circumstances? It is astonishing that the guiding principles are so far not sufficiently upheld neither in the chemical nor in the pharmaceutical industry. After all, we have safety rules to protect consumers in all other areas: cars and technical devices must satisfy legal norms or carry a certificate issued by some watchdog organisation.

In the case of chemicals and in the manufacturing of drugs and medical products such as dental restoration material, manufacturers and importers do not need to guarantee this kind of product safety. We may ask ourselves if there is any point to or use for our medical drugs laws. Why does it not pertain in the above cases? In the last few years, many naturopathic agents have been taken off the market, apparently because the high costs of certification do not justify their use. This is regrettable.

In chapter 7 we will encounter several examples of the havoc that can be wreaked by harmful substances: once a disease becomes irreversible, it is no longer curable and we as doctors can only look on. If prevention equals healing, it is our duty to point out these undesirable developments. If we continue to poison ourselves, our fate will necessarily mirror that of many of the plants and animals that continue to become extinct.

Plants are becoming extinct
In the press, we frequently read that hundreds of different species of plant become extinct each day, a fact that nobody seems to care about—at least nobody drew any serious consequences.

Marine animals are dying

A somewhat larger furore was caused by the dying of marine snails and oysters. The first of these phenomena to attract some attention were rationalized away as bizarre facts. When, at the end of the sixties, some scientists reported female marine snails developing male reproductive organs, this also was considered an anomaly. At the same time, French oyster breeders noticed oyster larvae dying without recognizable reason and adult animals developing balloon-like shell deformities.

The number of affected female snails increased in particular in the vicinity of the harbour. The cause was quickly found to be the chemical substance tributyltin (TBT), an organic tin compound that today still is the main ingredient of the paintwork of ships, in particular of large freighters. It is applied to prevent growths of algae, mussels, and barnacles on the ship's hull, as they slow down the vessel and cause it to consume more fuel.

The snails by themselves would probably not have caused any rapid reaction. Oysters, however, are of tremendous economical importance in the Mediterranean countries, where gourmets increasingly started to avoid them not only due to their lower amount of meat, but also due to their strange look. This lead to an immense financial loss and in 1982 the French government quickly reacted by banning recreational vessels from applying the toxic coating

TBT is only one example of the unimaginable effect chemicals can have on sensitive ecosystems and on human health. However, clearly no lessons were learned from these past errors, as demonstrated by the most recent press reports, dating from 2007, about the massive dying of bees in the USA.

> *"When bees disappear from the earth,*
> *humankind only has four years to live.*
> *No fertilization means no plants means no animals means no humans."*
> Albert Einstein

Bee mortality

In 2008 bee mortality rates started increasing in Germany and large parts of Europe. "First the bees die, then humans die", was on everyone's lips. Despite intensive research we still have not found the real causes of this massive bee mortality—either that or we refuse to believe them. The cause was said to be the Varroa mite, a bee parasite introduced from Asia. Asian bees are resistant against this pest, whereas the European honeybee is not. So far, this assumption has not yet been proven.

Beekeepers blame insecticides and pesticides—an assumption that seems obvious. In the past few years additional voices were raised blaming electromagnetic radiation (cell phones, mobile communication masts, electrosmog). Bees can sense the magnetic field, which is itself influenced by external magnetic radiation—a fact that also applies to pigeons.

Pigeons lose their sense of orientation—pigeon mortality
For more than ten years, pigeon breeders have been reporting that on some days, during competitions, their pigeons reached their home base either with a very long delay or even not at all. This phenomenon is called loss of orientation. Electromagnetic radiation once more presents itself as possible cause.

Pigeon breeders reported further shocking observations: newly-hatched animals died, but why? It turned out that the "black mats", manufactured in Asia and used as nesting material by numerous breeders for many years, were impregnated with the highly toxic TDI (toluene-2,4-diisocyanate) and MDI (4,4-methylene diphenyl diisocyanate). Both gases are used to prevent infections from being carried on intercontinental transports between Asia and other continents. It is however possible that chemical compounds with deadly side effects are formed in the process.

6.2. Human poisoning due to industrial accidents

However, humans are not spared. There exist acute and chronic poisonings due to harmful substances. Acute poisonings are easy to recognize due to the temporal and spatial correlation between disease and its cause. In contrast, for chronic environmental diseases any causal connections are hard to determine and prove. They develop in hiding, so to speak, sneakily, and creeping up on us over a period of years. Often there exist multiple causative agents until eventually one factor (a poison) causes the barrel to overflow.

Acute poisoning as a consequence of industrial accidents

Dioxins
Seveso, Italy, 10.07.1976: an aerosol cloud containing highly toxic dioxins escaped from a broken valve in the ICMESA chemical plant (Hoffman-La-Roche Ltd.). Post-accident, the number of infants born with deformities increased by a factor of 15.

Methyl isocyanate
Bhopal, India, 03./04.12.1984: more than 3,300 people died as a consequence of the so far most severe industrial accident in history, when 42 tons of the toxic gas methyl isocyanate escaped from the tanks in a pesticide factory operated by the US company Union Carbide.

Polychlorinated biphenyls
An industrial accident in Japan, in 1968, first demonstrated the polychlorinated biphenyls' (PCB) toxicity. In this accident, PCBs were able to contaminate rice oil through improperly sealed processing equipment, causing mass poisoning of more than 1,500 people. Similar accidents which led to the consumption of highly contaminated rice happened in 1976 in Korea (Ya Chon) and in 1979 in Taiwan (Taichong). The consequences—liver cancer, polyneuropathies, and hormonal disturbances leading to miscarriages and stillbirths—entered history under the moniker Yusho-disease. These accidents first stirred up public awareness of the PCB problem (93., 115., 116., 117.)

Chronic poisoning as a consequence of industrial accidents

Contamination of German drinking water (Ruhrgebiet 2006) due to highly toxic fluorosurfactants.
In 2006, the German media first reported contamination of the drinking water in the Arnsberg region with highly toxic industrial chemicals called fluorosurfactants. One year later it became clear that blood tests of 35,000 citizens, among them 350 first-graders, showed concentrations of perfluorooctanoic acid (a metabolite of fluorosurfactants) by factors of up to eight times the norm.

As usual, the authorities did not pay much attention to this incident, since it was considered impossible that these concentrations of fluorosurfactants lead to acute poisoning, and damage to human health was thought unlikely. These assumptions were based on two studies from the USA involving concentrations many times higher than in the Arnsberg case, which had not shown any health damage. We frequently hear this type of reasoning after incidents such as this. The fact that not only acute health damage but also the chronic consequential damage caused by these chemicals must be taken into account is overlooked or consciously hidden. Eventually, these acute accidents become a thing of the past and we can no longer obtain 100 % proof of causal connections—something which the legal system insists on.

How can it be that fluorosurfactants are classified as highly toxic chemicals but at the same time are being dismissed as harmless? It is true that the blood concentrations found in Arnsberg rarely cause acute poisoning. However, since their half-life is five to eight years, we can be sure of chronic damage to our health. We should inform the concerned parties of this and accordingly conduct long-term checkups. Maybe it is assumed that when, after ten to twenty years, the first symptoms arise, the victims no longer remember the incident. However, extensive study of the patient's medical history and testing their blood for harmful substances are capable of identifying these noxa at a later time.

6.3. Health risks due to nuclear accidents and temporary storage

- 1957 Kyshtym, Russia
- 1957 Sellafield (formerly known as Windscale), England
- 1979 Harrisburg, USA
- 1986 Chernobyl, Ukraine: operational error, security defects, and systemic weaknesses in the reactor core were causes of the so far most serious nuclear disaster during peacetime. More than 10,000 people died as a consequence of the accident and chronic diseases have developed in 125,000 of the 800,000 cleanup workers. (106.)
- 2011 Fukushima, Japan: caused by a massive submarine earthquake in the vicinity of the coast, and the ensuing tsunami. Three reactor cores were damaged. Two years after the catastrophe as many as 35 % of children living in the Fukushima prefecture (Fukushima Resident Health Management) displayed cysts and nodules in their thyroid

system—something which is normally very rare at this age. (68., 100.) One can only speculate about further consequences of the disaster.

As early as 1957, when I was still in school, I failed to understand why, in light of the two big nuclear meltdowns, we are not searching for alternatives, the more so since at the time there were no safe permanent or temporary storage facilities for nuclear waste. After Chernobyl all citizens of the European Union agreed that we had arrived at a turning point with regard to nuclear energy. However, no one asked the citizens for their opinion. Why are we not committing ourselves with zest to the search for alternatives.

Despite the numerous meltdowns that have happened, nuclear industry continues to maintain that nuclear power plants are safe. Confronted with this amount of ignorance, one would almost want to cry out:

What more has to happen to shake humanity awake and make it stop exploiting nature?

Studies found increased incidence of cancer in children who live in the vicinity of nuclear power plants. Despite several studies substantiating that claim, the official word is still that there exists no unambiguous connection between leukemia and nuclear power facilities.

Temporary storage facilities such as those located in Gorleben, Asse, and Morsleben are salt mines. As such, they are not suitable for storing radioactive material and are an additional imminent danger to our health—they are literally drowned in water. If this happens, radioactive material can no longer be reclaimed and if it can, it requires huge effort and poses a massive danger to public health. For Asse alone the costs are around four billion Euros, in today's money, and according to media reports, the process is estimated to take ten to fifteen years. As a consequence, groundwater is further contaminated with radioactive material and with it the creeping health damage continues, in particular genetic damage that is transmitted through generations.

Where now do we put those ramshackle barrels? This question is also still waiting for an answer. We need new temporary and permanent storage facilities. These are, however nowhere near being in sight. What applies to Asse is also valid for Gorleben and Morsleben. None of the responsible parties has the courage to concretely tackle the topic of where and how to store nuclear material. The costs of recovering the barrels of all three facilities are estimated at 10 to 12 billion Euros.

Wherever there are nuclear power plants, there are dangers to our health.

Despite the Fukushima incident, the new Japanese government wants to double check the decision to close all nuclear power plants by 2040. In my eyes, this is a grossly mistaken decision.

To summarize: there are dangers lurking in the winning of radioactive materials and inside the nuclear facilities themselves. Just as in the sixties in the USA and Canada, radioactive water is drained into rivers in the winning of nuclear material in Niger. Of course this is all disputed—there is practically no concrete evidence of any consequential damage. In addition, we can no longer estimate the risks posed to our lives by temporary and permanent storage facilities. If we are not even capable of creating secure storage in our own highly specialized country, what can we expect from the lesser developed countries such as Pakistan, Brazil, China, Russia, etc.

That which happened in the USA (Harrisburg), Ukraine (Chernobyl), and Japan (Fukushima) can repeat itself in our country, and in any other country, at any time.

6.4. Other harmful substances with creeping health effects
6.4.1. Drinking water and nutrition

Today already, our drinking water, and hence our food, are contaminated with pathogens, chemicals, heavy metals, and radioactive material (e.g., tritium). To this we can add hormonal degradation products, medical drugs, pesticides, solvents, etc.

As W. Soddemann (108.) reported in his arresting talk during the Environmental Conference in Hamburg (19.-21.10.2012), people, livestock, and wild animals excrete pathogens, which get released into the environment and the water supply. According to him, only 5,000 of around 20 million known chemicals are harmless, i.e., environmentally irrelevant.

Contaminated drinking water, meat scandal, free-range egg scandal, milk with carcinogenic fungal toxins—since the end of February 2013, we have read about them in our media on an almost daily basis. These highly carcinogenic substances are called aflatoxins. They are stable under heat and their presence in consumer milk is evidence that the state and its controlling bodies no longer have a firm grip on our nutrition.

What can one say when a country as rich as Germany is no longer capable to guarantee its citizens clean drinking water? Drinking water is the source of life. Without it, life cannot exist. A human being consists of more than 75 % water. It is obvious to anyone that water contamination will sooner or later lead to fatalities. Here we once more have to deal with a creeping danger.

6.4.2. Energy-saving lamps

Energy-saving lamps contain mercury, which is released when a lamp shatters on the floor and which can then enter our body via our breathing or via absorption through the skin, where it can cause massive neurotoxic, hormonal, and genetic damage. Besides mercury, lamp fittings contain further harmful substances that are dangerous to our health. Due to bad workmanship, in particular for bulbs that are manufactured in Asia (e.g., in China), these substances can be released at a steady rate. When we keep in mind the appeal made on 27.05.2011 by the council of Europe, importing these types of energy-saving lamps is irresponsible. (60.) Detractors say that the situation is exactly as it is wanted to be.

The following example shows what these bulbs are capable of. In Frankfurt, in 2011, a small boy was treated by an environmental physician after severe poisoning due to a broken energy-saving lamp. At the time, the boy was 4 years old. Two years later, in spring 2013, he is "merely" completely bald!

6.4.3. Contaminated soil, contaminated water, poisoned waste disposal sites

Frequently, our soil has accumulated radioactive uranium. The cause are mineral-containing phosphates, popular in agriculture as fertilizer, which contain uranium. This in turn enters the drinking water supply and hence the food chain. An even bigger danger to our health is posed by the big trans-regional waste disposal sites, where toxic substances from all categories (up to category III), coming from all over Europe, are stored without the local population being informed. It is a bit of a paradox that, as I observed myself, these sites are established in former nature reserves located at the border of towns, without appropriate sealing of the subsurface.

Figure 1. Two signposts pointing towards a waste disposal site located in a nature reserve.

Figure 1 shows that whoever put up these signposts was unaware of the explosive nature of the meaning and placement of those signs, from an environmental-medical point of view. Even worse is that none of the responsible parties seems to have noticed.

Figure 2. An 80 m high waste disposal heap with fissures (black line in the middle).

Figure 3. The same waste disposal heap shown in magnification. The black fissure now looks like a small brook.

Of course, waste disposal sites in such a close proximity to cities have consequences: the soil becomes contaminated and nothing will grow on it, the air is polluted with harmful substances to the extent that, depending on wind, one can smell it everywhere. The surface of streets, sidewalks and forest roads in local recreation areas are proveably contaminated with toxic metals, which we also find in snowmelt.

Figures 2 and 3 show fissures located in a waste disposal heap, which during rain turn into a brook which flows downhill. A large fraction of this penetrates the disposal heap and can go on to contaminate the groundwater.

Figure 4: fissures in a stone wall, caused by mining subsidence.

Figure 4 shows fissures that are clear indicators of mining subsidence, caused by a mine that was permanently closed in 2012. Rainwater clearly softened the foundational soil, which then led to it subsiding. It would be rather desirable to have a neutral authority to monitor waste disposal sites.

We can read in professional journals and DPA press releases about the increasing pollution of soil and air with toxic metals over the past five years (19., 20., 77.). This also affects plants and animals as our sources of nutrition.

Contamination of seawater

Here, I once more have to return to my observations in Africa. During my time in the deep african bush (1971–1974) I rarely if ever encountered the diseases of affluence that I mentioned earlier. I encountered some isolated cases when I was helping out in a large missionary hospital in the closest larger city Jos (population: 200,000). As late as 1981, when I revisited "my bush hospital", these diseases were still unknown there.

In contrast, when my wife and me revisited Africa (Burkina Faso and Tanzania), in 2005 and 2009, I found that there too our diseases of affluence are on the rise. Even worse is the fact that toxic substances, coming from the industrialized countries packaged in containers, are simply dumped into the ocean near the coasts of East and West Africa. The pollution of our oceans is programmed into the future, where we only stand and watch.

Already now we are seeing devastating consequences: massive fish mortality rates mean the end for fishermen. We can only guess at the effects of this all on the local population, so it is worthwhile to think about this. I hope to stimulate this kind of thinking with the information I provide, since I know these beautiful coastal beaches. One's heart starts to bleed when hearing about these issues in the media—when one has encountered them with one's own eyes and witnessed all these developments. It hurts, and one must scream out in pain—which I do, and I am not embarrassed about it. I owe it to the people of Africa, whose cheerfulness and wisdom I came to appreciate and love, to point all this out.

6.5. Electromagnetic radiation, mobile communications, electrosmog

The dangers to our health posed by mobile communications and electrosmog can already be sensed today. There are only estimate values for electromagnetic hypersensitivity because there is no obligation to report it—who on earth would it interest. According to these estimates, about 5 % to 10 % of people in Germany suffer from electromagnetic hypersensibility. The trend is a strong climbing.

Figure 5. New residential buildings, focussed in the reticule of power lines.

As many as 10 years ago, doctors, out of concern for the health of their fellow human beings, appealed to their colleagues, the general public, and the responsible parties in politics and in our healthcare system. Under the title "International Physician's Appeal 2012: Mobile Communications are Dangerous to our Health" (56.) they urgently pointed out the dangers posed by radio communication. It was translated into many different languages and was signed by more than 1,000 physicians, with support from a further 36,000 signatures worldwide. In the couple of years since, the signs of serious risks posed by mobile communication worldwide has multiplied and densified.[1]

---

1   Sources: Freiburg appeal 2002, www.freiburger-apell.info, 29.9.2012; Journal of the Society for Environment and Medicine (Umwelt-Medizin-Gesellschaft), modified on 4/2012.

Figure 6. Six new residential buildings, located underneath power lines

Figure 7. Under the slogan: "Property for your dream house at bargain prices" building now also has started on the left side of the street.

In their appeal, the undersigned write: "We as medical physicians have seen a marked increase in symptoms such as insomnia, chronic fatigue, headaches, migraines, vertigo, tinnitus, hyper- or hypotension and heart arrhythmias, concentration and memory lapses, learning and behavioural disorders in children, and an ever higher incidence of ADHD, in close spatial and temporal proximity to sources of electromagnetic radiation, in particular that caused by intensive cell phone use, DECT phones, WLAN, and in the vicinity of transmitting aerials."

In the meantime, many of the doctors' observations have been confirmed by numerous studies, conducted by independent scientists. The Freiburger appeal induced twelve further appeals by doctors reacting to these worrying developments, demanding long overdue preventative measures—so far to no avail.

Cell phones, but in general all types of electromagnetic radiation, seem to be especially dangerous to people who suffer from exposure to metals—at least, all of my patients with environmental health damage who complained about hypersensitivity to radiation showed years of exposure to metals. When transversing a space, the radiation emanated by the metals themselves interacts (interference) with other electromagnetic sources in the same area, a phenomenon called "coherency". Metals and radiation have much in common: they inhibit the activity of our immune system, reduce our body's defense cells, damage DNA and block its repair, stimulate tumor growth, cause metastasis[2] of existing cancers, and inhibit apoptosis[3]. As a consequence, their effect does not simply add up, but the two increase each other's effects, i.e., they triple or quadruple.

---

2   It might also be that these health damages, as well as cancer, only emerge through the interaction of metals and electromagnetic waves.
3   Apoptosis means intentional cell suicide, controlled by gene expression (the formation of genetic product, mainly proteins, encoded by a specific gene). In contrast to necrosis, apoptosis does not involve the release of cytoplasm and hence does not trigger an inflammatory response.

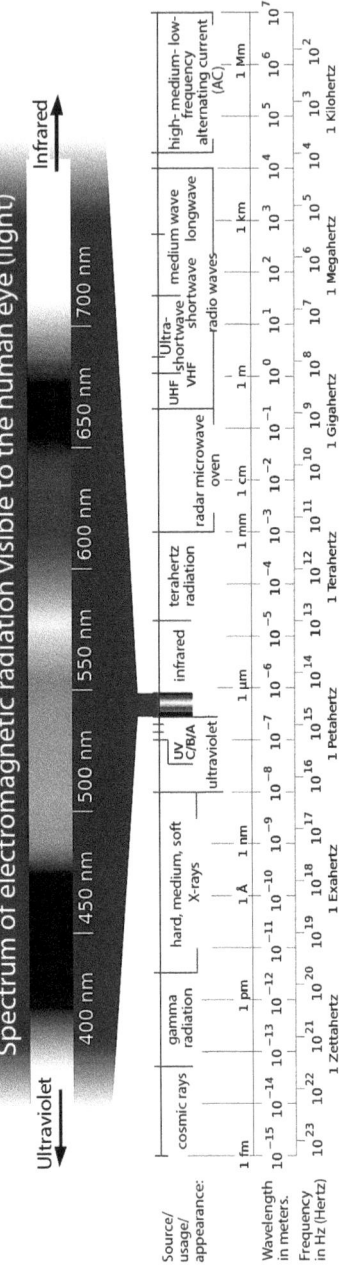

Figure 8. Frequencies and wavelengths of electromagnetic radiation.

The headline above a special issue of "Home + Health, Young People Research—and Once Again Find Something" reads: "Cell-phone-related Rouleaux Formation in Human Blood." (55.) The article goes on to inform us that "a 20 seconds' cell phone conversation is sufficient to change our blood count and cause coagulation of red blood cells. When we hold a cell phone to our ear, blood cells, which can usually move freely through the plasma and do not attach to each other, form rouleaux, start looking like frogspawn, attract each other as if by magnetism, and fasten themselves to each other, aggregating in lumps."

The aggregation of red blood cells then causes acidosis in the body's connective tissues that can no longer be compensated for by correct nutrition. In addition, intracellular blockages emerge so that orally administration nutrients cannot reach their target.

Our cells communicate through small electric discharges—for more detail, see Prof. F.A. Popp and M. Bischof's work on biophotons (10.). Hence the continuous interference from radiation at intensities millions of times larger, in connection with the blockages caused by metals is catastrophical in the truest sense[4]. Already now, we clearly see the grave consequences. I will return to biophotons in chapter 9.10. Science, as modern as it is, does not yet take them into account—no randomised trial, no double blind experiments to test drugs, vaccines, and medical products includes these biophotons, which are the light and hence the life of our cells.

New statistics issued by the Federal association of company health insurance funds (Bundesverband der Betriebskrankenkassen—BKK) show that the number of sick days due to burnout have increased by a factor of 18 between 2004 and 2011 (diagnose-FUNK [23.], journal of the Environmental consumers organisation for the protection against electromagnetic radiation). These are scarifying numbers. Nowadays, companies expect their employees to be available per cell phone without interruption. This flooding with information and working in a permanent flow of data elevate stress levels. It is not surprising that in light of the dramatic increase of burnout syndrome so many authors, such as B.K. Geuenich (39.) or K. Mül-

---

4   Bischoff studied Popp's work extensively and wrote his book "Biophotons, the Light Within our Cells" in Popp's spirit.

ler (86.), are concerned with this topic. Their work provides an impressive confirmation of the findings sketched above.

What are we facing? In October 2012, an italian court of law recognized a patient's brain tumor as an occupational disease. The tumor was caused by frequent cell phone use, holding long conversations during office hours. Hence, this radiation can also cause cancer. It is not known whether this patient at the same suffered from exposure to metals—this would be very important to know.

This case from Italy reminds me of one of my own patients, Ms. N. At the age of fifty-two she suddenly toppled over dead in her own home. Since her relatives refused a postmortem, we can only speculate about the causes. However, since I know her medical history, one probable cause clearly crystallizes in front of me. Ms. N. worked for a large travel agency in the capacity of sales representative. All day, she held phone conferences with hotels, bus companies, railroad companies, and airlines. She was always on the move and had to be reachable by phone at all times.

Hence, Ms. N. carried three cell phones, which she used for these work conversations. About a year before her death, I examined her for environmental harmful substances after she complained to me about headaches and psychiatric problems. My examination uncovered exposure to metals. Because she was close to a burnout, I urgently recommended her to take a sabbatical and, if possible, a change of jobs. Regrettably, she did not follow my advice, an advice that might have saved her life. Her early death deeply affected me. Should I have followed my intuition to conduct finer-grained tests? This had seemed impossible, because Ms. N. was not an inpatient.

In the meantime, the starting position for sound observational studies in the field of electromagnetic radiation is more than dubious. J. Mutter (89.) justifiably points out that "the question as to whether cell phones are dangerous or not can no longer be answered by scientific studies."

My advice is to test patients with hypersensitivity to cell phones and electrosmog for exposure to metals. If any is found, I recommend detoxification therapy in addition to avoiding the source of the radiation. The results of this approach are astonishing: the barrel that is full of toxins is partially emptied and the body once more becomes capable of finding its equilibrium.

6.5.1. Recommendations: what can we do?

About a year ago, the chairperson of a group that lobbied against the placement of a mobile communications transmitter came to me and asked what more they might do to prevent the placement of a transmitter[5] in the vicinity of our residential area. The group had lost all court cases they brought to trial.

After thinking it over for one week, I provided an answer: "Prepare a letter with approximately the following wording: dear Mr. N., with this writing we would like to inform you that we as residents will start documenting all our symptoms and diseases that appear from the day the building of the transmitter is finished. Should any anomalies in terms of frequency and severity of some diagnosis arise as compared to other districts of the town, we will sue you for many cases of negligent bodily injury, without prior notice."

The letter was signed by more than 200 residents. To this letter, they added scientific references of the danger to public health posed by such transmitters. After four weeks, the chairperson received an answer from the transmitter's operator, stating that he would relinquish the building.

My recommendation is to send such a letter at the start of the first round of discussions. This way, you can save yourself legal costs.

The failure to take preventative measures despite scientifically proven health risks posed by mobile communication is, in my eyes, no longer acceptable. There is overwhelming medical evidence for the danger that electromagnetic radiation poses to our health. It remains an open question as to the moral/ethical guilt of the responsible parties both from politics and from enterprise. Many of my patients, desperate in their hopeless situation, also asked themselves this question. Isn't this a case of deliberate failure to render assistance? Do courts consciously return incorrect verdicts in order to prevent a wave of new charges and the ensuing avalanche of legal cases? I am not at liberty to answer this question.

For this reason, I am thankful to W. Thiede (118.) who included ethical aspects into the study of the consequences of technology in his book "The

---

5   You can find the locations of all mobile communication transmitters under http://emf2.bundesnetzagentur.de

Myth of Mobile Communication: the Critique of Beaming Reason". With the help of our mobile technology and its derivatives such as TETRA or LTE, the myth of the individual's omnipresence and omnipotence appears to become true. In any case, economically independent studies issued warnings, since the so-called Reflex study proved damage to for instance DNA (double-strand breaks).

W. Thiede advises us to be vigilant. He concludes with the following: "In view of the seductive magic from our modern technology, we require ethically oriented demythologization." Not everything that can be done is ethically justifiable. In the course of this book we will encounter these borders more frequently.

6.6. The connection between infectious diseases and metals

Over the past 10 years, there was a rapid leap in the incidence of borreliosis[6], which is an infectious disease transmitted by ticks. All of my patients with borreliosis showed exposure to metal. We can assume that by now all of us are exposed to electromagnetic radiation to smaller or larger extent. Metals and electromagnetic radiation inhibit the activity of our immune system, which throws open the door to borreliosis. Treatment with antibiotics kills the bacteria. It is a fascinating question as to whether borrelia are the type of bacteria that store metals or prefer them as a source of nutrition, similar to bacteria and fungi in the bowel. If this is true, we must reckon with antibiotic treatment releasing metals, which in turn infiltrate the organs. I will discuss these connections in further detail in chapter 8.4, which discusses tin compounds.

All my borreliosis patients underwent detoxification treatment—without it, successful treatment of borreliosis is hardly possible.

6.7. Gene technology: curse or rescue?

Gene technology arrived at the scene to treat diseases in plants, animals, and humans. Although it is still in early stages of development, it already

---

[6] Borreliosis is the general name for various infections diseases caused by bacteria from the genus Borrelia (in the spirochete phylum). The most common carriers of these bacteria are ticks and lice.

slipped from our hands. Solutions are promised which can hardly be realized. The damage to public health posed by harmful substance progresses so rapidly that gene therapy won't be able to cure it neither in the near nor in the far future.

The handbook "Rare Diseases", published by Orphanet Germany in 2007 (94.) describes more than 1,800 rare diseases. In almost every case, we are dealing with diseases caused by genetic damage.

In the handbook's preface, Eva Köhler, at the time a patron of the alliance for rare chronic diseases (Allianz Chronischer Seltener Erkrankungen—ACHSE e. V.) writes: "In Germany, about four million people are affected by one of these ca. 7,000 diseases classified as 'rare'. Life is very tough for these patients." Of these rare diseases, 80 % are caused by genetic damage which is with high likelihood caused by exposure to harmful substances, as can be shown with biomonitoring.

In the meantime, this number of 7,000 diseases will be exceeded by a large amount. With this, I want to say that gene technology will always limp behind the number of patients affected by genetic damage. Hence, genetic engineering poses a danger the scale of which is hard to estimate.

Already today, the agricultural sector feels the consequences of misuse of genetic technology. In the US, Canada, Brazil, and several African countries there already are whole stretches of land where nothing wants to grow, after genetically manipulated soybeans or corn stalks were planted in its soil. As a consequence, neither weeds nor edible plants will grow there during the coming years.

> *Us humans will see the full extent of the negative consequences of our gene technology appear in our descendants.*
> *All of the responsible parties, however, refuse to believe this.*

In an open letter from 20.04.2011, addressed to District Administrator Dietrich Kühler, Gudrun Kaufmann (62.) points out these dangers to humans posed by agricultural gene technology. Ms. Kaufmann is a medically qualified health counsellor in the Odenwald special interest group for healthy living. Their topic is the massive health damage that can be caused by genetically modified feedstuff and plants, as well as the use of Roundup/Glyphosate.

Her statement boils down to the observation that epidemiological studies about exposure to Glyphosate show a correlation with the following severe health problems:

- miscarriages and stillbirths
- damage to the hormone system and to placenta cells
- multiple myeloma (a type of cancer)
- non-Hodgkin's lymphoma (another type of cancer)
- genetic damage

Hence, gene technology is another avenue for severe dangers to body and mind.

6.8. Conventional medicine at a crossroads—side-effects

Medical drugs and medical products in general are supposed to be for the good of people who are ill. The example of Thalidomide however shows us that all of these substances together with their additives (adjuvants) are foreign to our body and can pose damage to our health. This explains the increasing danger to our health posed by medical drugs over the past 20 to 30 years.

6.8.1. Medical drugs

Thalidomide, a sleeping aid introduced on 03.10.1957, was available over-the-counter. It was advertised with the slogan "Sleep and tranquility". The manufacturer pointed out its non-toxicity—it was safe even for children and pregnant women.

What followed next came as a shock to everyone attending the births of babies with missing or deformed extremities. Other had kidney damage, suffered from deafness or damage to the nervous system. After this wake-up call one would expect developments tending towards a decrease of side-effects caused by drugs. However, the opposite was the case.

In 1983, the renowned reference book of medical drugs, called "Bitter Pills", was published, authored by K. Langbein (70.). At that time, around 70,000 different medical drugs were manufactured by the pharmaceutical industry of the Federal Republic of Germany—an amount that even then was already unmanageable. Since then, there has been a further dramatic increase in the number of different drugs. "Bitter Pills" points out the necessity for medical drugs during emergencies, where they can save lives. At the same time, the book describes risks and side-effects of single drugs and of whole classes of drugs and readers are cautioned to be aware of these side-effects. What should one consider when taking a prescription drug? This is a question everybody should ask themselves in advance.

None of the usage guidelines and side-effect warnings were of any help.

In addition, many drugs are prescribed much too frequently and often many drugs are prescribed at the same time. Each medical specialist writes prescriptions that target the part of the body that falls under his responsibility and thereby overlooks any drugs prescribed by his colleagues. This leads to their accumulation in the body and, as a result, to them increasing each other's side-effects.

Hence, I was not surprised to read in the Westfalenpost newspaper from 23.10.2012 that drug cocktails cause sickness in senior citizen. The article says that "every tenth hospital admission of senior citizens is caused by a wrong mix of drugs. One third of people aged 75 to 85 takes more than eight medical drugs a day."

However, the topic of abuse of medical drugs does not stop here. On his blog *therapie.de*, Michael Malzahn cites Prof. Jürgen Fröhlich[7] as follows: "We assume that each year 85,000 patients treated in departments of internal medicine die of unwanted drug side-effects." Malzahn continues: "Prof. Fröhlich stresses that this number only includes casualties that are recorded in departments of internal medicine. The total number of casualties due to drug side-effects probably is much higher."

The Süddeutsche Zeitung newspaper also points out this fact, in their article "Prescription: Life". It reports that deaths due to drug side-effects are not reported and we may suspect a large estimated number of unrecorded cases. (33.)

According to the German Federal Office for drugs and medical products (Bundesamt für Arzneimittel und Medizinprodukte (BfArM), the number of deadly complications due to medical drugs is as low as 1,200 to 1,400.

---

7   Prof. Fröhlich leads the department of clinical pharmacology of the Hannover medical academy.

This number however only includes incidents that were reported. Once more, the truth is quite different.

In the Süddeutsche Zeitung newspaper issue from 16.07.2008, Claus Fritzsche writes:

> "Medical drugs are the third most common cause of death in Germany.
>
> In comparison, in the US, the number of drug side-effect-related fatalities totals 106,000, which makes it the fourth most common cause of death in the USA. In contrast to the estimates and extrapolations here in Germany, these numbers are based on scientific evaluations of several hundreds of thousands of medical histories."

When comparing Prof. Fröhlich's estimate of medical drug-related fatalities in Germany with those in the USA, we may conclude that the corresponding fatality rate here must be considerably higher than in the United States.

This large discrepancy between the number of fatalities here in Germany that are actually reported and the estimates and extrapolations also shows our need for significant public information. It seems obvious that neither the pharmaceutical industry nor the responsible parties within our healthcare system (in particular the BfArM) are interested in public education.

American studies on several hundreds of thousands of patients showed that only a third of all drugs has noticeable effects. Another third does not work at all and the rest has damaging side-effects.

However, the matter of side-effects not only concerns fatalities. Examples such as Thalidomide show how drugs can cause mental and physical damage leading to severe disabilities. In fact, package inserts warn against those dangers, although no lessons have been learned from this. As before, new medical drugs are in practice introduced to the market without appropriate neutral testing. The doctor finally prescribes these drugs in good faith of their tolerability. Even worse is that in the meantime many medications are manufactured in China and sold as generics, without appropriate testing. This means that we hand over our health to the Chinese who, if I may say so, are more interested in our money than in our health.

According to the daily news broadcast from 28.01.2013, studies conducted by the public health insurance companies also showed that simultaneous prescription of various drugs can lead to more side-effects than administration of a single drug. In my eyes, this is a very late realization—one which my boss pointed out to me back when I was still a junior doctor.

Since chronic diseases due to side-effects often only occur years after taking a drug, it is rather improbable that an attending physician will make a connection between drug and disease. This task becomes almost impossible once two, three, or even more drugs were prescribed. Maybe this is exactly the goal of multiple prescriptions: to make it impossible to prove a causal connection between disease and drugs or other noxa (harmful substances).

In summary, we may say that currently we are dealing with a cocktail of toxic substances, which consists of chemicals, metals, radiation (both radioactive and non-radioactive), mobile communications, UMTS, WLAN, TETRA, radar, microwaves, etc., and harmful drug residues. We should then not be surprised that, under these disastrous conditions, the number of chronic diseases and allergies has increased by more than 50 % to 100 %—sometimes even more. According to the Barmer GEK (a German public health insurance company), the incidence of for example major depression has increased by 117 %—something stressed time and again also by P. Jennrich (59., 60.).

One visible sign of this is the steadily increasing number of psychiatric and psychosomatic clinics. During the past decennia, the incidence of not just psychiatric diseases, but also dementia increased strongly and continues to rise dramatically. If I limit myself to looking back at the developments of the past two years, I expect a growth rate of as much as 100 % during the coming decennium. This fact may even apply to all so-called chronic diseases.

And what about cancer? According to M. Rath and A. Niedzwieck (100.), by now, cancer is the third most frequent cause of death in the industrialized countries. According to the World Health Organisation (WHO) 7.8 million people die of cancer each year. The number for only North America and Europe is 5.6 million, despite spending endless millions for therapeutic measures, in particular chemotherapy and radiation therapy.

To the pharmaceutical industry, patients with cancers and chronic diseases are a lucrative business that they'd rather not miss out on. An concrete example of today's prescription policies: a relative, a 90 year old lady, has been taking twelve different pills each day for months, amongst which are five drugs to lower her blood pressure and one pill, a diuretic, to support her kidney function. She takes additional six pills or drops when needed to remedy pain, constipation, insomnia, etc.

Currently, she has been suffering from allergic eczema for the eight weeks, which especially affects her face, back, and both her legs. In the package insert of six of these drugs we can read that the medication can trigger this eczema. Neither her family physician nor her dermatologist were prepared to wean her off some of these pills but instead prescribed her cortisone, another anti-allergic drug, and an anti-allergic salve. It was stated in the package insert that this medication may also cause allergic skin reaction in rare cases. Eventually, the eczema went into remission after I urged her attending physicians to remove the diuretic and replace it with a better-tolerated drug.

Aside from this, when establishing dose, doctors often do not take into account that women, children, senior citizens, pregnant women, and multimorbid patients often react much more sensitive to drugs than male participants in clinical trials that are otherwise in good health. The time factor plays an additional large role as drugs that are supposed to be taken over a period of months and years do more harm to our health than those that are only administered over a short timespan. This in turn never becomes clear from reading the package inserts.

In summary we can say that we should consider all medical drugs as substances foreign to our bodies and must appropriately inform patients about their side-effects. The same goes for vaccinations.

6.8.2. Medical products

In medical products, we can observe the same facts that we see in medical drugs. Here too, the danger to public health is steadily increasing, something which is, however, vehemently denied by manufacturers, distributors, and also by the responsible parties within the healthcare system. The best example is provided by amalgam, the story of which is quite representative for many other medical products as well as for medical drugs.

In her 1994 book "Ill from Amalgam", U. Hoffman (54.) empathically pointed out the risks posed by amalgam. She quotes several experts: "Throughout history, mercury probably has poisoned more people than any other substance. (Dr. M. Hanson) One day, we will determine that the unthinking introduction of amalgam into the teeth is one of humanity's biggest sins. (Prof. Dr. A. Stock) Sooner or later, amalgam always leads to damage to the nervous or immunity system and must be removed as soon as possible—preferably before any irreversible harm has been done. (Prof. Dr. M. Daunderer)"

I can only agree with all three experts. What applies to amalgam regrettably also applies to other dental restoration material. I will now say some words about the history of amalgam.

As S. Ziff reports (136.), amalgam was first introduced as dental restoration material in New York in 1826. Already at the time there were intense, sometimes highly emotionally charged, discussion. Since then, amalgam's story has been characterized by pros and contras, war (amalgam wars) and peace (hushups).

In the USA, amalgam was first banned in 1839, but was allowed back on the market in 1855. In Germany, the first ban was placed in 1926, thanks to Stock, a chemist, who urgently pointed out the material's toxicity in his numerous publications issued between 1926 and 1939 (110., 111., 112.). In 1928, the ban was lifted again. In his 1962 book "Poisons and Poisonings", J. Lewin (72.) also emphasized the toxic damage caused by mercury-containing amalgam. Ever since, amalgams have been in the focus of scientific investigation, as acting representative for mercury, so to speak.

In more recent times, even extensive publications by M. Daunderer (16., 17., 18.) and the scientific product monograph by J. Ruprecht (101.), director of the the scientific department of the Heyl concern, have not been able to silence Germany's amalgam's supporters. In Russia, amalgam has been banned for 25 years. Why is this not the case in Germany?

The Frankfurt amalgam trial, a long-time resident in the newspaper headlines in 1996, should set a trend towards the benefit of victims of environmental damage. The case is based on charges brought by ca. 1,500 private persons. As executive public prosecutor, Prof. E. Schöndorf (104.) headed the case of amalgam victims vs. amalgam manufacturers. Amalgam's opponents' hope was, however, in vain and in the end the trial was bent towards a result that agreed with amalgam's supporters.

The verdict was a bitter setback for the amalgam debate as a whole, and hence for environmental medicine. Despite convincing testimony by Prof. Wassermann *et. al.*, Degussa AG[8] was acquitted. His report was based on more than a thousand references to the scientific literature. At the time,

---

8   As of 2006, the company is called Evonik Degussa GmbH. It manufactures specialty chemicals and is particularly known in Germany as a manufacturer of amalgam and other dental alloys.

Prof. Wassermann (120.) was director of the Kiel toxicological institute. Led by the Institute of German Dentists (Institut der Deutschen Zahnärtzte—DZ), the dental profession took a stand against Prof. Wasserman's report and his reply (48. 49.). In my opinion, this is rather unusual.

The amalgam verdict states: "In the legal proceedings against [here, the three accused manufacturers are named] for bodily harm connected to the manufacture and distribution of dental restoration material (in particular amalgam), following § 153a of the code of criminal procedure and with approval from the court, we will at present desist public prosecution, on the condition that the accused pay, within one month, a sum of 1.2 million german marks. This sum will be used to fund scientific studies on the presence of heavy metals in the brain tissue taken from miscarriages and stillbirths, as a settlement with both plaintiffs."

When one studies this verdict closely, one must conclude that dentists skillfully and consciously conceal amalgam's toxicity. As an aggravating circumstance one must also know that that a large fraction of the patients involved in the trial were under treatment by a naturopath, a profession that in legal proceedings is taken even less seriously than environmental physicians. Apart from that, I expected at the time that no other verdict was possible within the politically predetermined boundary conditions—it was highly unlikely for Degussa to be sentenced, as none of the two parties involved was prepared to take responsibility for that.

Maybe however, it was not just a political decision. Another reason why amalgam's opponents and supporters came to a neutral compromise may be that at the time there were not yet any clinically convincing scientific results. In the end, claims of health damage due to amalgam, or other medical products or drugs, can only be proven or rejected by epidemiologic studies[9], which were not available.

Still, a string of private medical doctors and dentists was able to filter out a substantial amount of patients with possible health damage due to amalgam who could be healed for up to 80 % by treating them with chelating

---

9   In medical science, epidemiologic studies are conducted to determine the frequency, distribution, and causes of diseases. When causal proof of diseases is concerned, we speak of analytical epidemiology.

agents.[10] However, these findings only formed a so-called proof of general causality, i.e., they show a causal connection but did not provide medically and legally sound proof that mercury-containing amalgam was the single cause of the mentioned diseases.

In fact, with the benefit of hindsight, they were probably right. Besides mercury, amalgam also contains copper, silver, tin, and traces of nickel and palladium, all of which are highly toxic—as I will show in the following chapters. However, nickel and palladium must not be declared as constituents if their fraction is smaller than 0.1 %. When these other elements interact, they reinforce the damage each other does to our health.

Closing this "epidemiologic gap" is one more reason for my own studies, and for their publication in book form. I am very aware that my actions are perhaps only a small contribution towards the reappreciation of environmental medicine, but this is not my final goal. It is not the true goal that I personally try to reach, which is:

"Always follow your intuition! Intuition is the language of your soul."
I have lived after this motto for more than 60 years.

In the rest of this book, I will frequently return to the amalgam trial, since the verdict played a key role for environmental medicine. However, the verdict was a political/legal decision rather than one based on medicine, an assumption that appears to be confirmed by an open letter from Prof. E. Schöndorf (105.), sent on 10.06.1999 to the federal government and the members of the German parliament.

He writes:

"Dear madams and sirs,

please allow me to use this letter to call your attention to a problem that so far has not yet found sufficient importance within the sociopolitical discussion, but for which finding a solution is becoming ever more urgent. My argument is about the impairment of our public health by chemical substances. Without any doubt our chemical industry contributes to our economic prosperity thanks to its manufacturing of useful products and by providing jobs. However, it also affects us with its negative sides.

---

10 Chelating agents (from the greek root "chele" = lobster's pinchers) are substances that chemically clasp and surround other substances. This process creates a chemical bond with some component of the chelating agent clasping a metal.

More than 30 years ago, the example set by Thalidomide showed us the large risks connected with chemical-pharmaceutical progress. However, things were not limited to this individual case. In the 70s and 80s, many thousands of German people became ill due to toxic wood preservatives. These days, the problem posed by amalgam is intensely debated and a true apocalypse threatens on our horizon in the form of the collapse of our endocrine system and the breakdown of our immune systems caused by exposure to human-made chemical substances."

What happened during the following years ended up confirming the disregard in which politics holds environmental medicine. Toxicologic institutes such as those in Kiel and Dortmund are being closed. The subject of toxicology is now once more taught in institutes of hygiene and therefore made subordinate to medical science. Rather than adapting to new environmental conditions, developments were cut down.

We silently accept the fact that chemicals, drugs, and medical products that are supposed to serve the good of humanity cause physical disorders in the shape of chronic diseases and cancer. This continues the overall dying out of plants, animals, and humans as a consequence of poisonings.

Medicine must once more become human.
Over the last 10 to 20 years, the daily life of a doctor has increasingly been characterized by lack of time, pressure and requirements from higher-up, excessive bureaucracy, and economic pressure. The one universally valid motto is: our numbers must be in the black! This degrades patients and stifles their voice. There is no longer time for researching causes of diseases, let alone for planning true preventive action.

Hence, no-one will be surprised at the dramatic increase in the number of people suffering from chronic diseases, even though our aim is the opposite. Doctors should be able to take sufficient time for each individual patient. This can only be done by recording extensive medical history (anamnesis) and detailed clinical examination, both of which help create mutual trust. The patient senses that somebody is taking an interest in their case and cares for their health by going to the bottom of their case in order to remedy their disease. A chinese proverb that I learned during my training in acupuncture and TCM (traditional chinese medicine) teaches that:

> *"The best doctors prevent diseases and care about human health. Mediocre doctors care about not-yet-existing diseases and small-time doctors only treat existing diseases."*
> Chinese proverb.

Doesn't every doctor want to be an excellent one? Regrettably, reality provides us with a different picture. Yearly rewards go to hospitals that admit ever more patients and conduct ever more operations and transplants in ever shorter time. This is true for every hospital. The aforementioned universal motto has these clinics by the scruff of their neck. Under these circumstances, we wonder how the patient is supposed to regain their health? Does it even interest anyone?

6.8.3. Vaccinations

We cannot imagine contemporary medicine without vaccinations, which are necessary for retaining our health. In my capacity as specialist for tropical diseases and vaccinations, I spent 20 years as a consultant for the Westfalen/Lippe medical association in Münster. With regard to vaccinations, we must keep a few very important rules in mind, such as:

Administering two vaccines at the same time must be taboo.

When administering multiple vaccines, it is no longer possible to state with confidence which of the medications, or their additives, is the cause of any occurring side-effects, which substantiates the above claim.

After the chaos enacted by the vaccinations against swine flu, it has become clear to each citizen that we cannot really trust our experts also when it comes to matters of vaccination. At the time, the vaccine was not sufficiently tested. According to reports in the media, members of parliament as well as military personnel were given a different vaccine than regular citizens. The used vaccine contained thiomersal[11]. During September and October 2012, there was a shortage of vaccines to provide for the coming winter, which leads us to suspect that no vaccines *without* thiomersal remained. As to the reasons for that, we can only speculate.

It is incomprehensible that even today metals such as thiomersal and/or aluminium compounds (such as aluminium phosphate) continue to be in use as additives or as boosting agents. It is a known fact that both metals are highly toxic. Aluminium is a carcinogenic metal and mercury is a suspected carcinogen. Ethyl mercury in turn is a neurotoxin and many times more

---

11 In the body, thiomersal is metabolized into the highly toxic ethyl mercury.

toxic than inorganic mercury—a fact which I repeatedly pointed out in the past (130., 131., 134.). Because we do not know these properties for the additives that are in use today, we should refrain from their use.

I demand to refrain from adding metals to vaccines.

To many colleagues, the constituents of vaccines as well as their side effects are unknown, a fact explained to me by numerous patients. Patients are not sufficiently informed of this and parents in particular have not given informed consent—something which is required by law!

I stressed the dangers and side-effects of vaccines in numerous letters to the editors of local newspapers. The danger that flu vaccines pose to public health is made even larger by the calls, made through the media, for in particular high-risk patients (suffering from diseases such as chronic bronchitis or asthma, pregnant women, and people over 65 years of age) to let themselves be vaccinated. However, these high-risk patients must be examined particularly thoroughly prior to vaccinations and they must be fully informed of the risks. The attending physician should keep records of this.

With regard to vaccinations, I wish for a neutral supervisory body, one that is not held on a lifeline thrown from the pharmaceutical industry.

In Germany, vaccination is effectively required by law. Hence, the government should consult independent regulating bodies. Which side-effects can occur? Acute side-effects from vaccinations, such as fever attacks, reddening and swelling at the injection point, malaise, somnolence, are rare

As with all chronic diseases, the chronic side-effects of vaccinations only creep up on the patient over a period of months or years, at which point we can no longer determine their exact cause. In the past years, four cases came to my attention. After vaccination against the measles, two children went blind in both eyes and epileptic seizures developed in two other children. The worst about those cases is that the side-effects were not recognized timely enough, and the causal connection with the vaccinations was officially rejected.

Among the late consequences of vaccinations are epilepsy, multiple sclerosis, loss of sight, autism in children, and chronic rheumatic joint pains. In principle, when a vaccine contains metals and is injected into the body, any chronic disease can develop as a consequence. If the patient is already exposed to metals, e.g., during pregnancy, through intrauterine transmission from the mother to the embryo or foetus, the risk of developing an illness after vaccination is even higher due to the potentialization of side-effects (see chapters 7 and 8).

In the next two chapters, I would like to turn to the question as to what symptoms and diseases all these harmful substances (noxa) cause. In chapter 7 I will describe several cases of patients with an environmental disease. Chapter 8 will discuss the results of my extensive studies and the experiences that I gained in my practice for general and environmental medicine.

At the same time, I would like to encourage all my colleagues to scientifically evaluate their patient data. These days, this is much easier than before, thanks to the personal computer.

What do we take from this chapter? If we don't pay attention,
technology will replace morality and ethics.

6.9. Animal testing as pointless exploitation of nature

Now I will say a few words to animal testing. I find it irresponsible in our modern times to conduct animal testing for research purposes. This applies to medical drugs and to cosmetics. I say this consciously, even though many will brand me a moral crusader. Animal testing costs time and money—billions of euros even—even though there is no single confirmed case where human lives were saved. However, no consequences have been drawn.

Animals are exposed to large amounts of torture, even though we know that like humans, animals are highly sensitive towards pain. For example, metals are injected to find out if they can cause malignant tumors. Results from animal testing do not automatically carry over to humans, so why do we perform these tests at all? On these grounds, I oppose every single type of animal testing.

7. Case studies from my own practice

I wish for every one of my readers to at least once in their life experience the truly miraculous. During my pilgrimage, I encountered such experiences several times, which led me to my motto: "During a pilgrimage, we encounter no miracles, but the miraculous frequently happens." We can apply this motto to the field of environmental medicine, where I frequently saw proof of its truth. Clearly, one must experience for oneself how much appropriate detoxification therapy can achieve. He who convalesces, is right—something that I experienced time and again when working as an environmental physician. It helped me to continue consistently pursuing this area of medicine, in spite of all the backlashes and defeats.

Hence, real experience is more important than the written word. My previous supervisor, Prof. Dr. med. Kuhlmann (internal medicine), head physician of the Essen-Werden Protestant Hospital, frequently said to me: "Mr. Wortberg, when will you finally trust your own observations and experience." This is exactly what I also would like to advise you, dear reader.

As doctors, we are threatened by technology to have the experience and information obtained from the patients under our care taken away from us. Patients are silenced and are reduced to a number, a case, subjected to the laws of statistics and norm values (reference values)—something about which the people affected complained to me time and again in our discussions. It is not possible for the patient to defend themselves, because the medical establishment keeps them ignorant to a large extent, either consciously or subconsciously.

Medical action by doctors is determined by time pressure, equipment, and economical factors, which decide over health or disease. However, medicine—in particular environmental medicine—needs a lot of time, patience, and stamina, because each patient's story and problems are different. It is the task of the physician to fathom these reasons and remove them, no matter if you are a medical doctor, a naturopath, a homeopath, or practice holistic medicine.

A visit to an environmental physician often is a patient's final hope for cure, after many years of going from doctor to doctor without finding any help. I do not want to disappoint people who are as frustrated as these patients, so I give them my full attention and prepare them for their future in a way that is fully tuned to them as an individual human being.

7.1. Introduction: from my own life

Because I myself suffered from an environmental disease once, I would like to start by telling you some of my own life story.

I was very lucky to be able to travel to Africa as a development aid worker, rather than enter military service. My time was preceded by almost two-year-long scientific work in the biochemical department of the Marburg University pharmacologic institute. It was with a heavy heart that I refused the offer to qualify for professorship in pharmacology, extended to me by my Ph.D. supervisor, who was a wonderful boss, both professionally and on a human level. However, it was absolutely imperative for me to travel to Africa.

I prepared myself vigorously by working for three years in internal medicine wards, surgery departments, and gynaecology and obstetrics wards, followed by studying tropical medicine, hygiene, and environmental medicine in Liverpool (1966–1971).

Once in Africa, I worked for in a bush hospital in Akwanga (northern Nigeria) for three years. The hospital was a missionary clinic, run by Irish nuns. I was responsible for general, environmental, and tropical medicine, hygiene and surgery, urology, obstetrics, gynaecology, and pediatrics. I immediately noticed that of course there are numerous tropical diseases in Africa, but there are hardly any chronic diseases or diseases of affluence of the type that we encounter here in the West.

After I returned to Germany, I took training in acupuncture, TCM, homeopathy, and naturopathy whilst I was in the process of settling as a private doctor for general, environmental, and tropical medicine. I took my final examination on acupuncture and TCM at the age of sixty and almost always was the oldest student in any of the courses I took. In this respect, I may probably consider myself one of the few true allround-doctors—today referred to as holistic physicians.

That's it for now as concerns my medical career. My personal health problems proceeded as follows: as so often, it all began with my teeth. Before my graduation from high school, all students underwent a yearly dental check up. For most of the time, I hadn't had any issues, until a new dentist took over the examination in my final year. He pointed out that both upper incisors were still my baby teeth! This was a minor shock to me. Hence, after passing my exams, I decided to have gold crowns on both these baby teeth. According to my dentist they were made of "good gold" and for many years I tolerated the new material very well.

35 years later, I developed chronic sinusitis, which did not respond to drug treatment. Each morning, large amounts of phlegm came out of my mouth and nose that I could hardly catch with paper tissues. X-ray imaging then confirmed chronic sinusitis of the frontal sinuses (chronic sinusitis frontalis). It immediately became clear to me that my gold crowns may have caused the disease. Hence, I wanted to have them removed. I will forever remember the day it happened. When I woke up the next day, my head was as clear as it hadn't been for years. Maybe I didn't remember how it is to be clearheaded. Not a single droplet of phlegm came out of me and this hasn't happened ever since.

That morning, I traveled to Leipzig by car, to hold a lecture on tests I had conducted on protective trousers that I developed. It turned out to be a most pleasant journey! I was in high spirits and in great condition, and could have hugged the whole world. Ever since, I undergo detoxification treatment twice a year (chapter 10.3.1 and 10.3.3).

After my return from Leipzig, I sent my gold crowns to the Federal institute for material science (Bundesanstalt für Materialforschung—BAM) in Berlin to test the composition of the metals. Was my case one of bad luck, or not? A few weeks before I had my crowns removed, I had been lucky enough to visit BAM and attend a multiple-hour tour demonstrating the possibilities offered by electron microscopy. I spent four hours working with this miracle of technology, under the supervision of an engineer. With this type of microscope, one can look directly into the atoms and analyse metal alloys. I was so enthusiastic that I forgot that my wife was waiting for me in the car. I concretely wanted to know whether my gold crowns consisted of high-quality gold and platinum or whether it contained additional precious metals.

My tests found that the gold crowns to be not only gold-platinum alloy, but also contained palladium, of which elevated levels had been found during empty-stomach urinalysis. We have known since 1998 that palladium is highly toxic. These findings, and the way my healing progressed, prove to me that palladium with high likelihood caused my chronic sinusitis.

After detoxification, the gold-metal alloy was replaced with zirconium, which was tested for tolerability with the lymphocyte transformation test (LTT) before introducing them into my mouth. I have been in excellent health ever since.

7.2. Case studies

Now, I will sketch some examples of environmental diseases. I met all of the patients during their first consultation in my environmental health practice and they came from Lüdenscheid, the Märkischer Kreis, but even from northern and southern Germany, i.e., from as far away as 400 km. They wanted me to find out specifically if they were suffering from exposure to metals. I have guaranteed their anonymity by changing all names and dates. All patients involved, as well as the parents of the children I will describe, agreed with this publication.

The course each individual disease took is typical for environmental diseases. As attending family physician and expert witness, one suffers along with the patients. The whole extent of difficulties, the frequent denial of financial reimbursement from the health insurances (backed up by wrong or incomplete medical expert reports) and the negative verdicts from courts of law as regards disability compensation, all those affect both patient and their attending environmental physician in equal measures. Of course, the patient always is the weakest link. As an environmental physician, one always has several such patients, who cause a lot of work and worries.

Maybe some of my readers recognize themselves in one of these stories. These examples further show that cases like these cannot be comprehended with statistics and norm values.

All diagnoses stated after the patients' name and age are the ones made by the doctor who referred the patient to my practice.

7.2.1. The first spark: poisoning by dental restoration material

Richard, 45 years old: severe depression

Richard was a close acquaintance of mine. His medical history (anamnesis) shows the typical course taken by environmental diseases caused by metals released from dental restoration material. It progressed in multiple phases. Even though by then I had been working as a medical professional for twenty years, I never encountered such a story. When Richard came into my practice, in December 1987, he complained about many-years-long life of suffering. It all started with persistent headaches that afflicted him ever more severely over a course of three years.

Phase 1

None of the conventional medical examinations and treatments had been able to help Richard, which led him to consult a naturopath, who sent him to a dentist, due to the suspicion of amalgam fillings as the cause of his headaches. The dentist recommended dental restoration work, which he subsequently undertook. The dentist removed the amalgam fillings without proper precautions (dental dam[12], oxygen influx, specialized drill, etc.) and replaced them with cheap gold alloys that contained 30 % to 40 % more palladium. Richard did not undergo any detoxification treatment.

Phase 2

Richard's health temporarily improved, only to have his pain worsen one year later. He developed tachycardia. He became very short of breath and could no longer play tennis. Later on, he developed allergies to dust, pollen, and grasses. Richard became more and more depressed and suffered from back and joint pains. His depressive phases alternated with aggressive episodes. His marriage ran into a crysis. He went from doctor to doctor and from neurologist to psychiatrist. Eventually, the dentist decided to remove all the gold material and replace it with titanium implants. Once again, no detoxification was done.

---

12 Dental dam: the oral cavity and all teeth not affected by the dental work are covered with a plastic membrane which prevents mercury-containing fine dust from entering the oral cavity, from which it might enter the lungs or the brain directly.

Phase 3

Richard's condition worsened even further and his wife was extremely worried. When her husband had one of his aggressive episodes, he became increasingly violent, which made her afraid of him. He himself complained about permanent headaches, fatigue, poor performance, insomnia, problems concentrating, vertigo, loss of hearing, tinnitus, and dry eyes. He admitted that he was not able to influence his mercurial depressive phases and this was how he arrived in my practice, accompanied by his wife.

My clinical examination revealed minor loss of hearing in both ears. His blood pressure and pulse were slightly high, but there were no further pathological findings. Psychologically, he appeared a man who didn't know what to do and who lost all his vitality. He was completely dissatisfied with himself and his environment.

Ensuing urinalysis found exposure to metals, specifically mercury, palladium, tin, and organic tin compounds. All further laboratory values returned without pathological results. My main diagnosis was as follows: severe major depression with risk of suicide and risk of harming others, caused by heavy metal poisoning. To this were added the following secondary diagnoses: tinnitus, migraine (episodic, often pulsating headaches), first-degree loss of hearing, vertigo of unknown etiology, insomnia, and spinal syndrome—all of this as a consequence of heavy metal poisoning.

My therapeutic advice was renewed dental restoration work followed by detoxification therapy. This was however never done, since over the next few weeks, Richard started threatening his wife to the extent that I was forced to commit him to a hospital. Due to his severe depression and generally bad physical condition, renewed dental restoration work and ensuing detoxification were simply not possible. After a second compulsory hospitalization in a closed psychiatric ward, Richard did not appear in my practice, and I was unable to further track his ordeals. His wife obtained a divorce.

Richard B.'s example is significant for two reasons: first, it shows how important it is to obtain a thorough record of the patient's medical history, including a dental anamnesis. Second, this example immediately aroused my curiosity: could it be that amalgam fillings and dental alloys with their heavy metals are responsible for numerous diseases? Shouldn't we research this scientifically?

Where Richard B.'s case was the starting point for my environmental-medical studies, the medical history presented by an 82-year-old blind female patient (1979–2006) had already inspired me to conduct some scientific investigations. This patient lived in a retirement home and continually fell on her hips. The nurse in charge of her ward was extremely worried that she might break her femoral neck, which one day indeed happened, shortly after I had visited her. I answered the nurse's questions by saying that "if this patient falls again during the coming week and breaks her femoral neck, I will pull out some stops!" With this I meant that I wanted to scientifically study femoral fractures in elderly people. The cardinal question was whether a femoral neck fracture in the elderly is a consequence of a mild fall on the hips or, as was assumed until then, caused by osteoporosis.

I had managed to solve the problem posed by femoral neck fractures in the elderly only about one year before this case. In Travemünde, on 18.09.1986, I won the Max-Bürger award for an invention I made, the so-called "hip-pad" (also called protective trousers, impact neutralizer, or hip protector) and the proof of its efficacy. This award was issued by the German Society for gerontology and geriatrics. I summarized my work in a book called "Causes and Prevention of Femoral Neck Fractures in Elderly Patients. Presenting an Impact Neutralizer for the Prevention of Femoral Fractures in our Senior Citizens." (127.) This book was followed by two further publications on the topic. (128., 129.)

One year later, I wondered if the case of the elderly lady presented a new challenge lying in wait for me. When I was nine years old, I decided to go to Africa to help the people living there. It was the first major life decision that I ever made. To solve the question as to the cause of femoral neck fractures was the second major challenge.

In all of today's textbooks on surgery we can read that with 94 %, a minor fall on the hips (in the area of the great trochanter) is the major cause of femoral neck fractures in the elderly. Wearing a hip protector can lower this risk to around 90 %.

I formulated my new challenge as follows:

Are harmful substances (noxa), in particular metals and industrial products, responsible for the development of chronic diseases of unknown etiology?

Hence, Richard's ordeal became my motive for environmental-medically examining and treating all patients in whom I suspected poisoning by harmful substances and to analyze my findings. I meticulously carried daily journals for each patient, which was the beginning of a new, strenuous but exciting, part of my life. This time brought me many highs and lows, which I am hardly capable of putting into words. Once more, a portal into the environment, into the world, opened up in front of me.

One may wonder if all of this is coincidence or not. For me, there is no such thing as coincidence. I am a "number person", i.e., I try to back up everything with appropriate numbers. For 26 years, from 1979 to 2005, I tinkered with my hip-pad. Now, in my 26th year, I am once more turning to my book. After I received the award in 1986, I traveled all over Germany to give talks, followed by travels to the USA (Boston, 1992), Finland (Tampere, 1994), and Australia (Melbourne, 1996).

One highlight was my movie showing at the International Congress Center in Berlin (ICC), at a convention for German trauma surgeons. Because I, as a general physician, was not allowed to hold a lecture at such a renowned congress, I prepared a movie about my experiments with the protective trousers and showed it to the participants. In my movie, I showed falling experiments, both done 'live' by myself and in simulations, carried out in the wind tunnel operated by the Dornier company (an aircraft manufacturer) in Friedrichshafen. My talk was well-received and the surgeons were amazed. However, they did not support my efforts to have the costs carried by the public health insurance companies. It seems that they were afraid of a dramatic decrease in number of fitted endoprostheses. As health insurances and pension funds admitted openly to me during many conversations, this would imply that people get even older, which would no longer be financially sustainable.

So, now you know about me. Let us return to environmental medicine. In the following, I would like to present some case studies of children with environmental health damage due to intrauterine damage to the foetus. I will also describe my therapeutic strategy for each individual case (please refer to chapter 10 for details about medications).

7.2.2. Poisoning caused by intrauterine damage to the foetus

In all of the cases, the mothers wore amalgam fillings during pregnancy. During pregnancy, these metals were transmitted to the unborn child via the placenta and caused health damage to the embryo or foetus. None of the children that I examined had amalgam fillings themselves. In this chapter, I will only briefly discuss the therapeutic strategies that I applied.

Two examples of Ichthyosis (Ichthyosis congenita and vulgaris)

According to Pschyrembel (2012), Ichthyosis congenita is a rare disease that is either already present at birth or develops during the first few months of a child's life. It is characterized by cornification of the whole skin. Ichthyosis vulgaris is its most frequent and least severe form, which usually occurs during the first year of life, after the child is at least three months old. According to its definition, Ichthyosis congenita is hereditary, and the same applies to Ichthyosis vulgaris. However, if it turns out that a disease can be cured by detoxification treatments, we can no longer speak of a strictly hereditary disease. Patients have dry skin that looks like it has powdery scales.

Johanna, a neonate suffering from Ichthyosis congenita

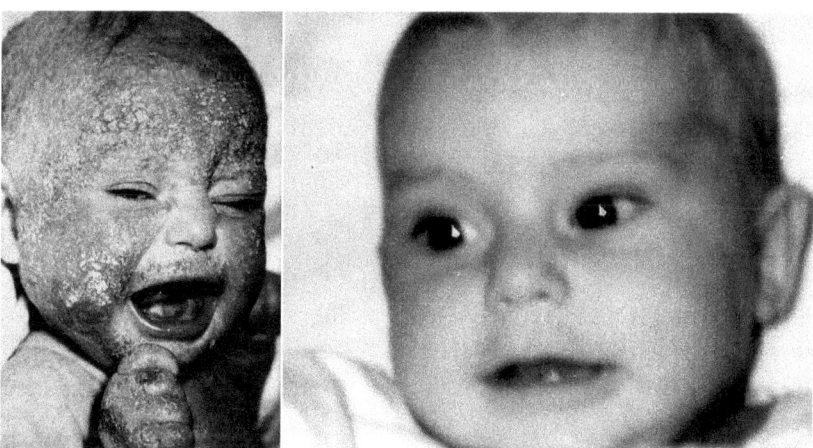

Figure 9 (left): neonate with Ichthyosis congenita, before treatment
Figure 10 (right): the same child after detoxification treatment.

Directly after her birth in October 1997, all of Johanna's skin was covered with scales. She was diagnosed with neurodermitis, but after several days it became clear that we were dealing with ichthyosis congenita. This was disastrous news for the parents, since it was commonly held these children could not be healed.

All of Johanna's skin, including her face, was covered with brownish, fish-like scales. Urinalysis showed exposure to mercury and I diagnosed her with Ichthyosis congenita caused by mercury poisoning due to intrauterine foetal damage.

I commenced detoxification treatment[13] with five DMPS (Dimaval)[14] injections.

Maria, 13 years old, suffering from Ichthyosis vulgaris.
Maria's parents told me about the ordeal their daughter had to live through: she was suffering from dry skin with powdery flakes, partially with thicker, dark-green to greenish scales. Here too, the initial diagnosis was neurodermitis. Then, her situation deteriorated and a University clinic diagnosed her with incurable Ichthyosis vulgaris.

How can one make such a bleak diagnosis? Conventional medicine's attention for the case faded away. Maria became restless and nervous and was afraid to leave her home, hiding in the attic. She could no longer wear sleeveless shirts: "That way everybody can see!" She hated sports, which forced her to undress and change clothes so that all children saw her skin rash. There was not a single general physician who knew what to do and the university hospital's children's ward was also unable to help her.

It was against this background that, in January 1998, her parents first visited my consultation hour. Maria was thirteen years old at the time. I found out that during pregnancy, Maria's mother wore many amalgam fillings and had part of them removed in this period. Empty-stomach urinalysis found exposure to tin and organic tin compounds.

This was another case of Ichthyosis, this time of the vulgaris variant. It was caused by amalgam toxicosis due to exposure to heavy metals, in the form of tin, and organic tin compounds.

---

13 The actual detoxification treatment was carried out by Dr. Wevers, from Wesel. At a later date, Johanna and her mother returned to me and reported about the treatment. Dr. Wevers kindly allowed me to publish this case report.

14 I will elaborate on DMPS and the DMPS test in chapter 9.8.2.

I gave Maria Dimaval capsules, zinc supplements, as well as vitamin E, and recommended her to drink lots of water. As little as three months later, the flaking of her skin improved. Maria quieted down and her performance in school improved significantly. She was in better spirits, went outside again, and later even participated in physical education classes. She has been healthy ever since.

Amalgam toxicosis in three children from a single family

The next three examples show that amalgam toxicosis due to intrauterine foetal damage cannot only cause diseases in a single child but can also affect several siblings. In March and April 1999, the mother and her three children first visited my environmental-medical practice.

Kevin, 14 years old, first child: headaches, prone to infections, hay fever, speech disorder.

First, I will report the case of Kevin. As an infant he was frequently ill and suffered from hay fever, headaches, and restlessness. As the mother told me, during the past three to four years he had become even more prone to infections and complained a lot about fatigue and vertigo.

Examinations by a pediatrician had not revealed any clear cause, and hence there hadn't been any improvements. He was also treated by naturopaths and homeopaths, once again without any results. His speech impediment could not be helped by any speech therapy, which eventually landed him into a school for special education. At the time no clinical examinations found any diseases other than his speech impediment.

However, empty-stomach urinalysis revealed exposure to copper, tin, and organic tin compounds (dimethyltin). My diagnosis was: headaches, prone to infections, hay fever, and speech disorder caused by amalgam toxicosis due to exposure to heavy metals, in the form of copper and tin, and organic tin compounds. The only cause that I could realistically consider was intrauterine foetal damage.

I treated him according to the well-tried protocol of Dimaval capsules and zinc orotate POS 20 in combination with microalgae. I additionally prescribed Solidago to support his kidney function and Hepatika drops to help his liver.

Two months later, his headaches had disappeared. One year later, Kevin no longer suffered from hay fever or any other infections. His fatigue had also disappeared, his performance in school increasingly improved, and his speech impediment disappeared.

Lisa, 11 years old, second child: eczema of unclear etiology, prone to infections.
Lisa's mother told me that her second child had been suffering from headaches and fatigue ever since she was six years old. Like her brother, Lisa also was highly prone to infections – if anyone with a cold came near her, she caught it too. The main reason for the visit to my practice was a permanently itching eczema that did not respond to any treatment and was a great cause of concern to the parents. Lisa was constantly scratching herself and was unable to concentrate for longer periods.

Treatment by a pediatrician had not had any results. All laboratory values were normal. Hence, they visited a naturopath and later also consulted a homeopath. Their treatment led to some small improvement, but neither of them was able to find the underlying cause.

Lisa's empty-stomach urinalysis showed exposure to the same substances as were found in her brother Kevin, namely copper, tin, and dimethyltin. Here too, we made the diagnosis: headaches, eczema due to amalgam toxicosis, both caused by exposure to copper, tin, and dimethyltin.

I treated Lisa the same as I did with her brother. Once again, the situation improved significantly three months after detoxification treatment. The eczema went into remission, headaches disappeared almost completely. Lisa became calmer and more balanced, and her performance in school improved markedly.

Nadine, 19 years old, third child: headaches, prior gallbladder surgery, therapy-resistant eczema of unknown etiology, prone to infections
Like her siblings, Nadine suffered from headaches, inner turmoil, fatigue, was prone to infection, and had eczema that did not respond to treatment. Furthermore, I noted that her gallbladder was removed at age 18, due to gallstones. None of the conventional medical laboratory tests found anything pathological. On various occasions, Nadine consulted naturopaths and homeopaths who also were unable to offer anything more than temporary relief. None of them achieved permanent improvement.

Urinalysis showed exposure to metals in the form of copper, tin, and dimethyltin. Like her siblings, we treated her with Dimaval capsules, zinc orotate POS 20 tablets, and Solidago and Hepatica drops. After only a few months, her headaches improved significantly. Nadine became allover more tranquil. Her performance in school improved and she was no longer prone to infections.

For all three children, detoxification therapy had been successful. Hence, we must assume that exposure to metals had caused their diseases. These cases are an example of how easy environmental medicine can be.

The laboratory tests and the ensuing therapeutic successes in all three cases show that metals can be transmitted from mother to foetus via the placenta, where it can go on to cause toxic damage leading to diseases later in life. We get the impression that the incidence and severity of the symptoms increase with age, as indicated by Kevin's speech disorder and Nadine's gallstones.

Charlotte, 17 years old: migraine
These days, children no longer suffer from just headaches. Ever more frequently children visit my environmental practice reporting full-blown migraine attacks, often at ages as young as five to ten years old. I would now like to describe one example.

When Charlotte consulted me in July 2000, she did not at all appear sick. She seemed in high spirits and she vividly described her symptoms. Her father confirmed that she was a good student, despite her health problems. Despite numerous visits to doctors, the underlying cause of her migraine attacks had not been found.

Besides headaches accompanied by nausea, Charlotte also suffered concentration problems. The headaches sometimes persisted for as long as two to three days, forcing her to stay in a dark room, unable to do her homeworks. There was, however, a temporary correlation between her symptoms and her period. During the past two years the migraine attacks had piled up. Charlotte's mother had worn amalgam fillings since childhood, i.e., also during pregnancy.

Once again, we carried out empty-stomach urinalysis, which found exposure to copper, tin, and organic tin compounds (dimethyltin). My diagnosis: migraine due to amalgam toxicosis caused by intrauterine foetal damage.

Administering Dimaval, zinc, and vitamin E effected significant improvement. The severity and frequency of her headaches decreased. Years later,

I met her father again, who reported that Charlotte was completely symptom-free and is now a successful university student.

AD(H)D (Attention Deficit and Hyperactivity Disorder)

As of today, the cause of ADD and ADHD is supposedly unknown to conventional medicine. As it is a syndrome it designates, according to Pschyrembel 2012, a group of symptoms characteristic for a specific disease (phenotype) usually with consistent causes (etiology) but unknown history (pathogenesis). I.e., their pathophysiological genesis is unknown. The incidence of ADHD is becoming ever larger and requires a lot of patience and empathy from both parents and teachers.

I would now like to present some examples:

Dennis, 14 years old: ADHD

Dennis is another case that I will not forget anytime soon. Despite the efforts of all of his family, including parents and grandparents, this boy's upbringing was a never-ending story, which led them to send him to a boarding school six months after first starting primary education.

When they first visited me, in November 1999, Dennis' mother reported concentration problems and hyperactivity, and the boy had been incontinent up to age eleven. His performance in school was mediocre, despite his above-average intelligence. Aside from this, he displayed markedly abnormal behaviour. His attending pediatricians diagnosed him with ADHD—in the past these children were referred to as 'fidgety kids'—and prescribed Ritalin, which he however did not tolerate. Because his parents did not see any improvements, they consulted me in my capacity as environmental physician.

During my examination, Dennis was constantly in motion. He was swaying his legs and frequently stood up and walked around touching some object or other. For a fourteen-year-old boy this is rather anomalous behaviour.

Urinalysis showed exposure to mercury and copper. The diagnosis was once again obvious: ADHD due to intrauterine foetal damage caused by mercury and copper poisoning (amalgam toxicosis). Since the disease had clearly affected the boy's central nervous system, I commenced detoxification with DMSA (dimercaptosuccinic acid), followed by a change of regimen to Beta-Reu-Rella (microalgae) and Paracilantro (coriander herb). As little as three months later we saw a marked improvement and therapy continued, with some interruptions, for three years.

At the age of 19, Dennis visited my practice and assured me that he is now healthy and wants to train as a carpenter. He came to me, because he wanted to thank me for my efforts. This was another day that I remember vividly—it is quite uncommon for a young man to visit his attending physician years later, just to thank him. I am sure that Dennis will find and follow his road through life.

A child with severe cerebral damage, caused by exposure to copper

Intrauterine foetal damage can lead to severe damage to the infant's brain and can hence cause problems for mental and physical development of the child. The earlier during pregnancy that the noxa attack the embryo or the foetus, the more severe the damage done. I would now like to report one example. To my knowledge, no similar case has been described before.

This case exemplifies the chasm between conventional and environmental medicine that exists today. Maybe it will cause some wonderment and shaking of heads, but it should stimulate thinking and help start the renewed discussion about metal poisoning.

Timo, three years old: early childhood brain damage with mental and physical disability, of unclear etiology. Disease allegedly incurable. Additionally: muscular hypotonia[15].

Timo's parents, grandparents, and Timo himself coped with his disease with a lot of courage and love. Friends of mine told them about my environmental-medical practice. When she entered my practice, in December 2002, Timo's mother carried her three-year-old son on her arms. He was a large child but could neither walk nor speak and only emitted incomprehensible sounds. Later-on, he crawled on the floor like a seven-months-old baby. His father was 40 years old and his mother 32 years old. Both of them were in good health. During pregnancy, his mother had had a few amalgam fillings replaced.

When Timo was born, he seemed to be in good health, with a birthweight of 3,500 grams. However, only a few months after his birth, significant developmental disorders appeared, which led his parents to consult every single pediatrician in their surroundings.

---

15 Reduced muscular tension when in motion, caused by CNS disorders and disorders in the spinal cord.

The conclusions drawn by prominent neurologists and human geneticists: early childhood brain damage with mental and physical disability, of unclear etiology. Additionally, he was diagnosed with muscular hypotonia. His parents were told that his symptoms could be treated, but there would be no specific therapy to cure Timo's disease. They must anticipate Timo never learning to walk and speak properly. Since he was nine months old, he frequently underwent physiotherapy to treat his mental and physical developmental defects and the pronounced weakness of his back musculature. He regularly attended the center for human genetics for check-up tests.

During my clinical examination I noted that Timo was very tall for his age being closer in dimensions to a five- or six-year-old. I didn't find anything pathological when examining his heart, lungs, abdomen, and skin. His arm and leg reflexes also showed nothing unusual. Timo was unable to speak and generally seemed very slow.

When we were playing hide-and-seek and I crouched behind my chair, he displayed some curiosity. He furtively looked at me from the opposite side of the chair. He also was very interested in my glasses, which he tried to take off my nose. However, he appeared mostly uninterested in his environment. His mother told me that Timo still had to be fed by his parents and was not yet potty-trained. Mentally, he made a friendly and quiet impression on me although his severe physical and mental impediments were impossible to ignore. Timo most closely resembled a very young infant.

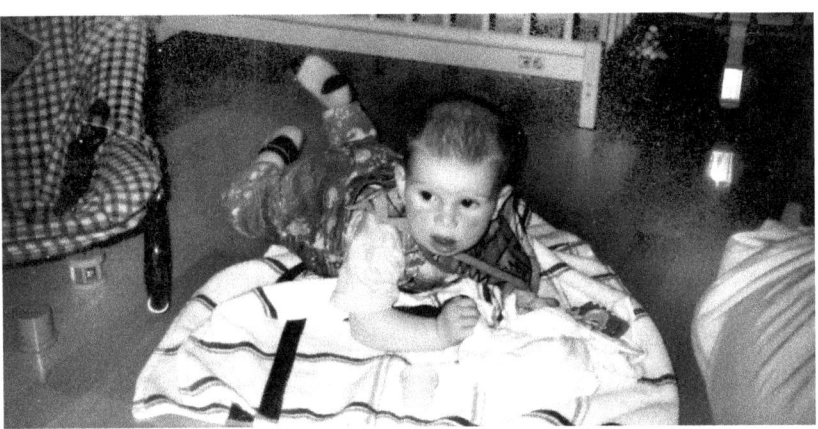

Figure 11: Timo, 3 years old, suffering from severe mental and physical impediments, before treatment.

Four months after he first started physiotherapy, Timo still was unable to sit or stand on all fours. He was, however, capable of moving his head slightly. His eye movements when tracking other movements were very brief. When grasping objects, he touched them only briefly and with a lot of caution. Hence, he no longer wanted to tear my glasses off my face. However, I noted that at the time he could at least crawl around with zest.

When Timo's disability was diagnosed, his parents' world collapsed. They were not willing to accept it and visited me to ask if Timo's plight might have been caused by exposure to harmful substances. None of the laboratory tests (blood count, liver enzymes, blood sugar, thyroid hormones, and pituitary gland), conducted in numerous clinics, found anything pathological. Hence, we did not conduct these same tests again.

Since Timo's mother had four amalgam fillings remaining, we examined his empty-stomach urine for presence of heavy metals contained in amalgam, i.e., mercury, copper, and tin. We did not determine levels of silver, since their insurance company only paid for testing of the first three metals. I found mercury and tin levels within accepted norms. With 357 µg, however, copper was elevated by a factor of 60 times the norm value (5 to 50 µg/l).

I diagnosed Timo with early childhood brain damage with severe mental and physical disability and muscular weakness, caused by heavy metal poisoning with copper, incurred during pregnancy. Both parents wore amalgams. Empty-stomach urinalysis also showed copper exposure in Timo's father, whereas his mother did not any longer show any exposure at that time.

Since Timo did not wear any amalgam fillings, his exposure must have been transmitted through the mother. During pregnancy, copper was transmitted through the placenta into the foetus (see chapter 8.3 and 9.7).

I commenced detoxification treatment with Dimaval (2,3-dimercapto-1-propanesulfonic acid as sodium salt) (16., 101., 103.). Dimaval is a so-called chelating agent. It contains SH-groups that easily bind with heavy metals. SH stands for a so-called Sulfhydryl group (= sulphur group). Due to these SH-groups, Dimaval displays a high affinity to many sulphur-containing heavy metals and binds with them to form stable complexes.

In addition, I prescribed zinc supplements (zinc orotate POS 20), vitamins, vitamins, and Solidago and Hepanest drops from the Nestmann company.

I recommended the following therapeutic protocol:

- Dimaval, one capsule each 3rd to 5th day, depending on tolerability.
- zinc orotate 20 mg, one capsule daily around 18:00, which is the time of day when zinc orotate is best absorbed by the body
- vitamins C and E in doses appropriate to children
- Solidago N drops to support kidney function, starting with 3x5 drops and increasing to a maximum of 3x15 drops
- Hepanest to support liver function, starting with 3x5 drops and increasing to a maximum of 3x15 drops
- Paracilantro drops[16] to detoxify the brain.

Three weeks later, we replaced Dimaval with DMSA. DMSA, a chelating agent, has the advantage of being able to pass the blood-brain barrier and can hence also remove metals from the brain. Timo tolerated DMSA very well to the extent that his body almost demanded it: whenever his constitution weakened, his mother knew it was time to administer DMSA.

Two more months later, Timo's mother reported that her son somehow had become increasingly curious about his environment. He was opening drawers and closet doors, and took an interest in his environment and he was in better shape allover. However, his parents also needed to look after him more closely. He was now able to walk without assistance. Within months, the baby had turned into a toddler—this remarkable development was completely unexpected to Timo's parents and family.

Check-up empty-stomach urinalysis six months later only showed mildly elevated copper levels of 58 µg/l. As before, mercury and tin were within acceptable norms.

---

16 Paracilantro consists of coriander herb extract which, after D. Klinghardt, can pass the blood-brain barrier, binds to heavy metals, and then removes them from the brain.

After seven months, his mother reported that her son's bodily constitution had improved even further. He could now stand up straight and they could go for a two-hour stroll—something that was impossible, inconceivable even, before treatment. He is now also more friendly and more "low-maintenance". His mental constitution was, however, not improved in any substantial amount, although he now on occasion tried imitating words.

At age six, Timo started to attend kindergarten, where he was largely accepted by the other children. He was now able to say a few words, such as "mommy" or "daddy", but could not say these words on demand. His main improvement was that he was now able to entertain himself and no longer needed strict supervision. This was a big relief, in particular for his mother and grandmother.

I visited Timo when he was seven years old, in order to observe him in his domestic environment. His mother, grandmother, godmother, and his now already 17-year-old sister Sofia, were present at the occasion. After his parents had followed my recommendation to undergo dental restoration work and ensuing detoxification, they had a wish for another child, and got a healthy daughter. Timo himself appeared cheerful and healthy, aside from several mental and physical limitations.

Timo's family members looked after him with love and devotion. When I entered their garden, he greeted me with a hug, took my hand and pulled me through the gate onto the sidewalk and ran away, with me in tow. He clearly was eager to show me that he was now able to run. To be able to do this like other children was a major step forward to him. He got along well with his sister.

Timo liked going to school (a school for special education), progressed steadily, and was in high spirits. He already was able to draw several letters, copying them from examples.

Figure 12: Timo at age six, together with his healthy sister, after detoxification therapy. He can now walk, ride a bicycle, and speak a few words such as "mommy" and "daddy".

Figure 13: Timo at age six, riding a bicycle (tandem).

Figure 14: Timo at age six, riding a bicycle all by himself.

One year later, I visited Timo once more. His parents reported further progress: they were all very surprised and at the same time very happy when all of a sudden he called his father "Manfred", his mother "Beate", and his sister "Baby". However, he rarely if ever imitated others' speech. His parents worried when, one day, he burnt himself but did not display any signs of having pain. They were happy that he could now go skiing and was able to swim without needing assistance, activities that he really enjoyed. By now he was also able to ride his bicycle all by himself, as long as someone rode behind him.

We then went into his room together, where Timo immediately jumped on his swing, made from linen cloth. He started swinging, but he did not sit on it but rather suspended himself in the cloth. During past visits to my practice I always had the impression that he was not able to focus his eyes, but rather stared through things. When I visited him at home, I noticed that he could now observe objects and people, albeit only for a short period of time. This too was a major improvement.

Figure 15: Timo at age seven, riding a horse.

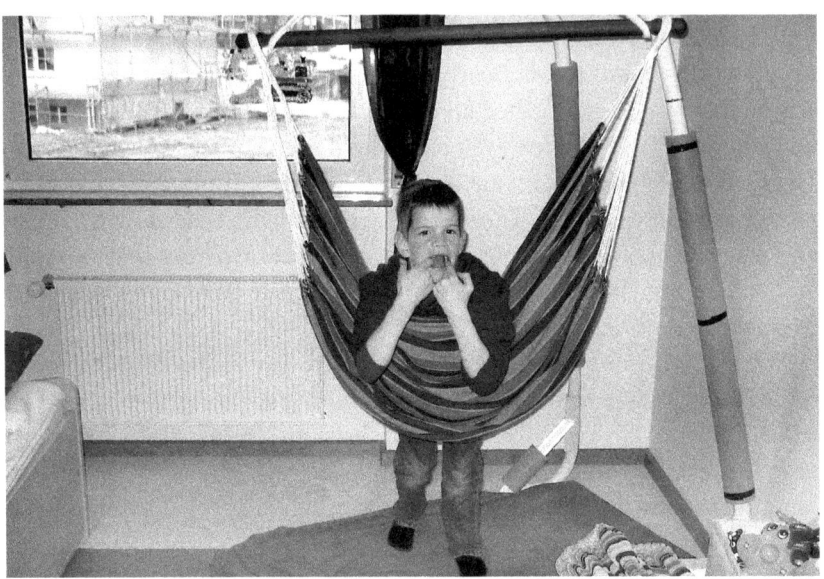

Figure 16: Timo at age seven, playing by himself.

Figure 17: Timo at age seven, eating and drinking unassisted.

Figure 18: Timo goes to school and learns writing.

Timo continued to follow his therapeutic regimen and regularly took pills to help detoxification. By now, his mother inserts a two-to-three-week break every three to four months, as agreed between the two of us.

At age eleven, Timo, his parents, and his young, healthy, sister moved to Japan where, according to his mother, children with disabilities receive much better care than in Germany. He now attends secondary school. His detoxification therapy continues.

What do these examples of intrauterine foetal damage show us?
When speaking of heavy metal exposure and health damage due to dental restoration material, we always first think about mercury in amalgam and/or about palladium, that is an additive in gold- or palladium-containing alloys. These case studies testify that other metals, such as copper and tin, and organic mercury and tin compounds also can be transmitted from the mother to the foetus, causing severe, irreversible health damage when the child is still a baby. Concretely, Timo was exposed to copper within his mother's uterus. His example also goes to show that, after dental restoration work and appropriate detoxification, mothers can definitely carry healthy children.

In the scientific literature, there are hardly any cases of diseases caused by organic metal compounds in humans. For more on this, please refer to chapter 3.9.

7.2.3. An example of expert witness testimony

Matthias L., 18 years old: malignant brain tumor.
Preparing environmental-medical expert testimony is a massive challenge for every environmental physician. It is frequently connected with disappointments, culminating frustrated patients and doctors. The reasons for this are obvious.

In Germany, as in most industrialized nations, environmental medicine plays hardly any significant role: its proponents are often laughed at, despite the availability of excellent scientific publications on the topic. This conscious ignoring is reflected in jurisprudence, since judges usually follow the recommendations made by expert witnesses connected to universities and medical academies. I would now like to report one of these wrong assessments.

The case concerns the legal dispute between Matthias, 18 years at the time of the trial, and the city, whom he sued for damages after multiple-year exposure to PCBs during his time in primary school. He had developed a malignant tumor, which had been operated on at a later stage. He sued the city and requested compensation.

The court sent me a written request[17], summoning me to prepare an expert opinion. The plaintiff's complaints should be substantiated by evidence of a causal connection between his diagnosed malignant brain tumor and his exposure to polychlorinated biphenyls (PCBs) emitted in the classrooms of the primary school that he attended between the first and fifth grades.

Although many years have passed now, Matthias' story and the legal proceedings sound hard to believe. During the case, Matthias was represented by his mother. Since she could not find any university professor to prepare the report, she asked me to write an expert opinion. After brief consideration, I agreed, in full knowledge of the bleak prospects of legal success.

I was pessimistic, because during a previous case, the judge stated to me during pretrial that on the one hand, I had convinced him 100 % in the medical sense, but on the other hand he was only 99 % convinced, from a legal point of view, that dental restoration material had been the cause of the plaintiff's symptoms. Hence, the main trial reached a verdict in favour of the health insurance company. The plaintiff, who was most severely ill, withdrew her charges. The reasoning given in the verdict came as such a shock to her that her blood pressure rose dramatically. Two days later, she suffered cerebral haemorrhage and a stroke. She has been almost completely (level 3) invalid ever since.

With this case in the back of my mind, I nevertheless stated my willingness to prepare the report.

In the following, I will intentionally describe the case in some more detail. It is intended to be an example of how legal proceedings can develop. The fact that Matthias' mother had made a diary of her son's disease made it a lot easier to determine its course over the years.

Matthias first visited my practice in October 2004, at an age of 18 years. When he was 12 years old, his brain tumor had been removed. Hence, there was a gap of six years between surgery and his first consultation in my practice—this fact explains the added difficulty in establishing proof. The surgery was performed in a university clinic and ten days post-op he was transferred to his local hospital.

---

17 I will reproduce this request in slightly modified form, since its legalese is incomprehensible to the layman.

Family anamnesis

His family had no history of serious diseases. In particular there were no known tumors or blood diseases.

Matthias' individual history

Matthias was the second child born to his parents. His birthweight was 3,060 grams at 51 cm. At a later age he always was particularly tall for his age.

When he was transferred from the university clinic to his local hospital, at age 12, he was already 169 cm tall and weighed 75 kg. Aside from some pediatric diseases and a domestic accident, his development up to starting school had been inconspicuous.

Social anamnesis

Matthias' parents originally came from Rumania and divorced before Matthias was born. There was no contact with the father, which made his mother, Ms. L., a single parent. She raised him and his sister. His older sister is more than twelve years his senior and she is in perfect health.

At the time of my expert testimony, Matthias was 18 years old. There was great mutual trust between mother and son. His mother was is a certified translator of the Russian and Bulgarian language, but retrained as a geriatric nurse. She always had concerns about her son—as a geriatric nurse she almost exclusively worked night shifts, leaving her more time to spend on him.

Mother and son traveled a lot, both in Germany and abroad. Matthias always coped well with these travels and he had always been in good health.

Educational anamnesis

Matthias started primary school at age six. He was initially required to visit preschool, but then started the first grade. When he attended the second grade, he first began to complain of concentration problems and minor problems with reading. During his time in primary school, he had a remarkably high frequency of illness and complained of nausea, vomiting, and stomach pain, after returning home from school. His family physician always maintained that these are school-related nervous problems. Twice, Matthias visited a pediatric clinic, but nothing that merits attention was found.

Due to his problems in school, it was decided that Matthias should go to a school for special education. This was, however, strictly rejected by his mother, since she was convinced that Matthias is equally gifted as his fellow students. She then visited a pediatric psychologist, who made him take a test. On the basis of its results the psychologist certified Matthias to be slightly more gifted than other students in his age group. When Ms. L. showed these results to the headmistress of the school for special education and sketched the whole situation, the latter then pronounced him fit to stay in primary school.

Symptoms

Somewhere around the end of Matthias' first year in school and the start of his second, the first symptoms appeared. He complained about problems concentrating and some minor problems with reading. The symptoms worsened during the year, and stomach pains and nausea also developed. During that year he had been ill quite frequently. Whenever he was at home during the weekend, or traveling, he felt perfectly well. The symptoms worsened during the second and third years, and he also started complaining about headaches.

Due to his frequent illness, he missed a lot of classes and fell behind in school. His hair started falling out and his fingernails looked freckled, which incited his mother to take him to their family physician. He diagnosed him with local hair loss (Alopecia areata) and psoriasis of the finger nails. Both diagnoses were later confirmed by a dermatologist. Time and again, the family physician traced back his problems to school-related nervous issues, but his mother was not prepared to accept this.

Due to his frequent absence from school and his bad performance, Matthias was put back in second grade, where Ms. B. became his new teacher, who remained with him during third and fourth grade. Ms. B. and Ms. L. had lots of conversations, as the first also felt negatively affected by the time spent in the classrooms. Ms. B. often asked her, "Do you smell anything strange in this classroom? I always smell something. When I get home, I've got migraines and teary eyes."

Whenever Matthias came home from school, he was very tired and needed to lie down. He did not want to play outside. This continued all through his third year in school. According to him he felt gobsmacked and weak. He had always had an interest in biology and microscopy and wanted to become a doctor. However, from the third grade onwards, he also lost interest in this, which was a big worry for his mother.

When he was in the third grade, laboratory tests were first conducted on the air in the classrooms, probably incited by Ms. B. These tests brought no results. During the fourth grade, Matthias started to suffer from increased itchiness. Other children now also started complaining about headaches. Once again, upon the suggestion of Ms. B., air samples were taken from the classrooms and tested in the laboratory. Ms. B. was not at liberty to discuss the results.

That same year, Matthias went to hospital for an EEG. No reasons for his complaints were found and he was discharged with the diagnosis "healthy".

The following year, Matthias and his family traveled to Austria for a holiday in the mountains. Here, he first started complaining about dizziness, vertigo, and double vision. It was the first time that his complaints persisted when on holiday. After returning home, their family doctor ordered antibiotics. Since there was no improvement, he was admitted to a pediatric clinic in his hometown as an in-patient, on 19.01.1998. After three days, he was discharged with the suspected diagnosis "migraine".

According to his mother, no magnetic resonance tomography (MRT) examinations or EEG were made during his stay in the hospital. In my opinion this should have been done at that time. However, his mother was not entirely sure, because not all findings had been discussed with her. Back home, Matthias first said out loud that he is seriously ill and his mother took him to their family physician that very same day (22.01.1998).

This doctor ordered an urgently necessary neurological examination, which however did not result in any pathological findings. His EEG showed wave patterns characteristic for a concussion. It was decided that the therapy ordered by the family physician should be continued.

On 24.01.1998, Matthias started complaining about strong pains in his eyes. His mother then contacted a cousin in Poland, who worked as a doctor. She recommended her to take Matthias to an ophthalmologist, who determined elevated pressure on the optic nerve and sent the two of them to an eye clinic right away. They arrived in the evening, where computer tomography imaging of his skull was performed. The head physician then showed them the images and told them the shocking news that Matthias had a brain tumor that was pressuring the optic nerve. The tumor had to be operated on as soon as possible but, because it was located deep inside the brain, it was very difficult to operate. Their best chances were to go to a department for pediatric neurosurgery in a specialized clinic. At this point in time, Matthias already suffered from vertigo and shortness of breath.

His mother contacted the neurologist, who was prepared to make a phone call to the specialized clinic, which stated their willingness to immediately examine Matthias and, if necessary, perform surgery. Since their public health insurance would not approve paying for an ambulance service to the specialized clinic, located 350 km away from their home town, at such short notice, his mother drove him to the university clinic with her own car. As recommended by the attending physicians, they administered cortisone before and during their journey. Matthias cannot remember the drive—he must have lost consciousness whilst on the road. In any case, once they arrived in the specialized pediatric clinic, he was unconscious. Only when they were in the elevator, he briefly opened his eyes.

The very same day MRT (magnetic resonance tomography) imaging of his skull was performed with the diagnosis:

- strong suspicion of a malignant brain tumor
- acute hydrocephalus (hydrocephalus occlusus)

Matthias was operated on in the neurosurgery department that very same evening. He had been lucky one more time: his tumor was completely removed via microscopic surgery and the water that had accumulated in his brain was let out via a drain.

Pathologic-histologic testing found "malignant germinal cell tumor of mixed type". His further progress was nondescript. After ten days, Matthias was discharged from the neurosurgical clinic and transported to a pediatric clinic in his hometown, to continue his treatment, where he underwent chemotherapy and radiation treatment. As of today, Matthias visits the clinic for after-treatment.

Postoperative school career
Despite the seriousness of his illness, Matthias qualified for fifth grade of middle school. He received tutoring in English, German, and mathematics, which continued after his dismissal from hospital.

Matthias was confined to a wheelchair for five months. During this period he did not qualify for financial support, leaving caring for him solely in the hands of his mother. Because of this, she was not capable of working full-time and, with her being a single parent, it left them into dire financial straits.

His chemotherapy caused Matthias to lose 27 kg, which he later gained back. For the first four years after radiation treatment, he suffered regular anxiety attacks and had problems sleeping.

Starting with sixth grade, he attended middle school. He was a good student and gained a certificate stating him to be capable of handling the pressures of Gymnasium education, which he looks forward to starting.

Clinical diagnosis during first visit

Matthias is a tall, broad-shouldered eighteen-year-old, who is slightly overweight (92.5 kg standing 188 cm tall). He seems alert and quickly and correctly answers every question that is asked. His general condition and nutritional status are adequate. He displays slight redness and flaking near the hairline on his neck. A scar in the median section of his head has healed in exemplary fashion. There are no pathological organic findings.

Psychological diagnosis

Matthias displays no psychological abnormalities. He appears to have come through the major surgery and the ensuing chemo- and radiation therapy without complications. He intends to take control of his own life, which expresses itself in his insistence to spend a year in a foreign country.

Laboratory findings

I have largely skipped classical laboratory tests, since these are conducted by the pediatric clinic at regular intervals and showed no significant pathological postoperative findings.

Findings from environmental-medicinal laboratory tests

The lymphocyte transformation test (LTT) for chemicals and metals showed no pathologies, other than a questionable hypersensitivity to nickel.

The tests for exposure to metal and/or harmful chemicals did not show any pathological results. After six years, finding exposure to chemicals in the blood was no longer to be expected.

Main diagnosis: postoperative condition after removal of a malignant germinal cell brain tumor of mixed type.

Polychlorinated biphenyls (PCBs)

The PCB accidents that occurred in Japan, Korea, and Taiwan brought the dangers to public health posed by PCBs to attention once more. After these accidents, PCB was gradually forbidden by all industrialized nations. In 1978, the Federal Republic of Germany (former Western Germany, BRD), limited use of PCB due to its toxic properties (47.). It was only allowed in closed systems such as electric transformers, hydraulic oils, and capacitors. In 1989, a complete ban on PCBs was introduced. The Bavarian ministry of health, nutrition, and consumer protection prepared a summary about PCBs (5.).

PCBs can be released into interior spaces through leaky, PCB-containing capacitors used in fluorescent lamps, and also through radiation from walls and expansion joints due to it being used in silicon. (4., 109.) PCBs also escape from engines used in household appliances, office machines, and heating pumps.

How high are the threshold values? The former federal health authority recommends a "preventative threshold" of 300 ng/m^3 total PCB in interior air and an "intervention threshold" of 3000 ng/m^3—this means that when this second value is exceeded, action must be taken and schools as well as kindergartens must be closed. The source of danger must immediately be removed.

Toxicokinetics and Toxicology of PCBs

PCBs are related to dioxins, have a similar mechanism of actions and are often contaminated with each other. PCBs deposit in practically every single organ. They are potent neurotoxins and highly toxic to the brain and the liver. They damage the immune system, cross the placenta border, cause deformities and are carcinogenic. According to BUND, we also find PCBs in breast milk (15.).

Measurements in the classrooms in Matthias' school revealed PCB levels as high as 26,000 ng/m^3. Matthias had been exposed to these levels of emission for five to six years—a very long time for any child, but also for the general population. To this was added the fact that Matthias was overweight. As early as age eleven or twelve, he had the height and weight of an adult. According to M. Teufel (115., 116., 117.), PCBs like to deposit themselves in adipose tissue. This also applied to Matthias.

Because of PCBs carcinogenicity, and in light of Matthias' prior history, we must assume them to be responsible for his developing the germ cell tumor within his central nervous system (CNS). At least the course his illness took and his symptoms suggest this. What are some typical symptoms of a germ cell tumor in the brain?

- double vision
- headaches
- vomiting on an empty stomach
- disturbance of vision
- decreased level of awareness

These signs are identical to those expressed by Matthias. Other reasons for assuming a causal connection between PCBs and his brain tumor were two further fatalities in school employees due to CNS tumors that occurred during the aforementioned period. Other students and teachers also complained about headaches, burning eyes, vomiting, nausea, and unpleasant smells. Elevated PCB levels were found in the blood of two other teachers, as told to me by Matthias' mother during our anamnesis. However, these results were not supposed to be revealed publicly out of privacy concerns.

It is likely that the elevated PCB levels in the classrooms of Matthias' primary school are not just caused by emissions from fluorescent lamps but also from sealing joints and/or concrete walls. Hence, renewed measurements are necessary, depending on the degree of renovation of the building.

Audit report

According to the most recent inspection report, technical defects in the tubing were found in 32 of the 84 inspected rooms. At the same time, indoor measurements of PCB levels were conducted, after pressure from a female teacher. The results, however, have not been released and were requested and published only in the context of the legal proceedings.

My answer to the expert witness question

I have answered the posed question as follows: from a medical point of view there is a high likelihood (more than 90 %) of a causal connection between the plaintiff's developing a malign germ cell tumor and the emission of Polychlorinated Biphenyls (PCB), to which Matthias L. was exposed during the time he spent in the classrooms of the primary school supervised by the defendant.

This conclusion is backed up by the following facts:

1. The defendant's medical history.
2. PCB is classified as a toxic and carcinogenic substance.
3. In animal tests (monkeys and rats), PCB's carcinogenicity has been scientifically proven beyond any doubt.
4. With high probability, the plaintiff was exposed to PCB during the six years he attended primary school, where above-threshold PCB concentrations of more than 26,000 ng/m^3 were measured.
5. As a child, the plaintiff belonged to a high-risk group, for whom the concept of threshold values does not apply with respect to health.
6. Due to his height and being overweight, Matthias was at an increased risk.
7. In addition to Matthias' case, malign central nervous system (CNS) tumors were found in two school employees. Furthermore, students, teachers, and other employees displayed similar symptoms.

The plaintiff's case was dismissed by the Higher Regional Court. The defendant, in its role as regulatory authority, was pronounced not guilty and was found to not have neglected its regulatory duty.

This ruling is completely beyond comprehension given the pressing burden of proof provided by this expert witness' review and the listed technical defects found in the school, as well as the omission of PCB measurements in the classrooms. The municipal government was cleared of all charges. Matthias' mother did not receive any kind of compensation. My expert testimony was ignored to the extent that it was not even mentioned. During the trial, the causal connection between brain tumors and PCB exposure were not discussed at all. I was not asked for a verbal expert witness statement, even though this is common practice in legal cases such as this.

The court explicitly prohibited any further appeal. The plaintiff's counsel accepted the ruling and refused further dialogue with me, the expert witness. It must however be added that the plaintiff's initial counsel suffered from serious illness that started during the legal proceedings and had to hand over the case to a new legal representative, with whom I have had no contact other than an introduction by phone.

This example shows that also from a legal point of view, environmental medicine only plays a minor part in this country, especially when the defendants are public authorities and/or public health insurance companies, pension funds, or trade associations. These institutions seem to have carte blanche for their dealings. Is this supposed to continue for ever?

Matthias' case is of some relevance from a legal point of view, because any wrongful verdict sets a precedent for similar cases. Even today, we encounter many cases of this happening.

In order to understand the verdict, we should know that from 1964 to 1980, PCB was in use in the building sector and in fluorescent lamps. During this time, the post-war reconstruction program built 46,000 new schools and numerous other public buildings such as kindergartens and city halls. Hence, PCBs possibly affected many thousands of public buildings. No one suspected at the time they were erected that toxic substances in the form of PCB might hide in these sometimes very tall buildings.

Both mother and son have never been able to understand the verdict, which is more or less a tragedy to them. As expert witness, I consider it a wrongful verdict, aimed to prevent claims for compensations from other parties against cities and municipalities. This assumption is backed up by the fact that Matthias' malignant tumor was more or less accepted as a fact of life, even though scientific proof of causal connections was available. The current state of science was simply ignored.

In their WESTPOL broadcast from 09.09.2012, the German public broadcaster WDR reported about PCB in schools. The report explained that it is not Nordrhein-Westfalen public building ministry that is responsible for maintaining their buildings in such a way that there is no risk to public health. Rather, the regional building regulations stipulate that it is the municipalities' responsibility, which includes staying below legal thresholds for PCBs. The report went on to reveal that there is, however, no compulsion and no legal directive to regularly test schools for presence of harmful substances. Measurements and renovation only proceed if there are concrete grounds for doing so. So, it turns out that milk must be spilt before action is taken.

Hence there exists a legal gap even today. Frequent control measurements are only required if it is proven that PCBs were strictly used in building during the relevant period. This is in turn impossible because there frequently are no records of this. According to the broadcast, municipalities abuse this legal loophole.

This leads us to the following questions:

1. Is this fact morally and ethically justifiable today?
2. Who is responsible for this legal loophole?
3. When will it be closed?

It is not my task to provide an answer to these question. It is the duty of the responsible parties.

In any case, Matthias went to the USA for one year as an exchange student. He then belatedly took his examinations and after graduation successfully studied business economics. I only saw him again many years later when, one day, he visited me, bringing with him a binder filled with documents and newspaper clippings pertaining to his court case. When he left, he turned around and said, "Mr. Wortberg, may I give you a hug? You are the only human being who helped me and fought for me." I was left speechless and could only say "thank you".

To me, this case taught me an important medical lesson: it is highly likely that PCBs can pass the blood-brain barrier, something which at the time was not publicly known or at least was denied by the responsible parties. One reason for this effect may be that constant exposure to PCBs weakens the immune system, thus opening up the blood-brain barrier. A diseased patient is subject to his own laws.

7.2.4. Depression and asthma: reimbursing costs of medically required dental restoration work is denied

Rita N., 45 years old: depression and asthma bronchiale

Ms. N. visited me in September 2009 diagnosed with depression and asthma bronchiale. She consulted me in my capacity as environmental physician, because the public health insurance company refused to pay for medically required dental restoration work. She told me that her diseases were caused by dental restoration material. It all started with elevated blood pressure, headaches, and tinnitus. She no longer had any capacity for working under strain and was always tired, having an increasingly hard time coping with her work as kindergarten teacher. She stressed time and again that she was in fact a very active person. Later, laboratory tests found hypothyroidism and a nodule in her thyroid. She received medications, which she did not tolerate well.

Upon recommendation by the naturopath, the amalgam filling was removed without any of the necessary precautions and subsequently replaced with gold inlays and synthetic material. At first, the situation improved, but after as little as one month the patient's health deteriorated. She started to have fever attacks and swollen lymph nodes, after which she developed bronchial asthma. In addition, Ms. N. suffered increasingly frequent depressive episodes and was certified incapacitated for work, which was particularly harsh for her, since she specifically loved practicing her profession.

Both urinalysis and blood tests showed hypersensitivity to palladium and nickel, previously classified as Type IV chronic allergy. I made the following diagnosis: bronchial asthma, depression, and fever attacks with swelling of the lymph nodes, as a consequence of heavy metal toxicosis. Secondary diagnosis, also attributable to poisoning: high blood pressure (hypertonia), hand eczema, thyroid adenoma with accompanying hypothyroidism, tinnitus, migraine.

After removing the new dental prostheses and ensuing detoxification, the situation clearly improved, only for a new problem to appear: the health insurance company refused to pay its partial contribution to this new dental restoration work, despite my three separate, scientifically founded expert reports on the case.

The expert opinion issued by the MDK (German medical service for the public health insurance funds) found no causal connection between the patient's disease and the materials used for the dental restoration, despite correlation between the time when the restoration work was done and the appearance of symptoms, and despite the demonstrated exposure and hypersensitivity to heavy metals as well as the fact that, after the second restoration work and ensuing detoxification, the patient's situation improved to the point where she was able to practice her profession again.

From an environmental-medical point of view, both of the MDK's expert opinions were found lacking. The MDK did not take my scientifically founded objections, as stated in my own expert opinions, into account. It should be pointed out that, as they stressed to me during many discussions, in such cases the insurance companies are more or less forced to follow a negative decision of the MDK.

After the MDK's third rejecting opinion, I issued a fourth one, in which I concluded that I felt forced to file a complaint for medical misassessment with ensuing bodily harm against the MDK's medical experts, unless they prepared a scientifically founded counter opinion. Once more, no contrary opinion was issued but instead the patient and myself received a demand to submit documents that we had already submitted months ago. Within a few weeks, the patient received notification that the costs for the second restoration work would be carried by the insurance up to the maximum amount allowed by the relevant laws.

This case provides a typical example of how public health insurance companies and experts from MDK treat patients who suffer from environmental-medical health problems.

When cases escalate into legal proceedings, new medical experts working for the insurance companies usually issue similar negative opinions that do not contain any scientifically justified statements. Therefore, when one encounters obviously ill-prepared expert opinions, it is recommended sue the MDK and the insurance company directly, rather than let the case go to court.

This is clearly an uncommon strategy, but at least it is logical and, hence, honest towards the patient. It is regrettable that the MDK applies such methods, which are, incidentally, usually very disappointing for the employees of the health insurance companies, who largely see through this "game" the MDK likes to play—at least, during almost every discussion I had with said employees they admitted this fact after I pointed it out to them.

7.2.5. Huntington's disease

Kurt K., 40 years old.
When one first meets a person suffering from massive tremors, it is hard to believe that it is possible to live with such a condition. When Mr. K. and his wife entered my practice, in October 2001, I was faced with a man who hit around himself with his arms and legs in rapid but irregular intervals—that

is, he stretched his extremities in all directions in a highly uncoordinated manner.

Mr. K. had been a bus driver for 10 years, thus spending every day on the roads of his city. Previously, he had worked at a metal processing plant for fourteen years. Since he found the work to be too exhausting and too dirty, he retrained to be a bus driver. About five years ago, he first noticed occasionally losing control of his arms: they didn't do anything other than just wildly flapping around him. In the course of time, they started shaking without him being able to control it in any way. A neurologist eventually diagnosed him with Huntington's disease. None of the therapies tried nor any of the medications or physiotherapy helped in any way. This strongly affected Mr. K. and made him in part depressed, in part aggressive. His children in particular were scared of him, but his wife also often felt threatened, and their marriage was in danger of falling to pieces.

Aside from the tremor, clinical examinations showed no pathological results. Psychologically, Mr. K. seemed very somber, sad, and dispirited, like a man already close to giving up. He was not able to explain his aggressive phases, at least he was not capable of controlling them.

Mr. K. told me that, initially, he only had amalgam fillings in his teeth, which were removed without proper precautions approximately six years ago to be replaced with gold crowns. No detoxification was carried out. In light of the fact that the law mandates that, due to their toxicity, amalgam fillings have to be disposed of under highly controlled conditions, with punitive measures upon violating this law, it is impossible to understand how the procedure above could have been performed.

Blood tests then showed a questionable hypersensitivity[18] to gold and inorganic mercury. Usually, given that it's questionable, this does not merit treatment, but the patient should be monitored. None of the other general and environmental-medical laboratory tests showed any pathological results.

---

18 A lymphocyte transformation test (LTT) was used to show the hypersensitivity in the blood. The measured parameter was the Type IV cellular hypersensitivity. In the text, we discuss the stimulation index (SI).

An SI > 3 corresponds to a more than threefold activation compared to baseline and shows the existence of allergen-specific T cells in the patient's blood (positive result), i.e., cellular hypersensitivity.

An SI < 2 counts as a definite negative result.

An SI between 2 and 3 should be considered borderline (weak, questionable hypersensitivity) and monitoring the patient should be considered.

There was no doubt that Mr. K. suffered from Huntington's disease and major depression. In this case, the interesting question was to what extent the questionable hypersensitivity to mercury and gold could be considered as causing the patient's disease.

Although none of the conventional and environmental-medical laboratory tests showed any pathological results, no reference values were exceeded, and only the questionable hypersensitivity to mercury and gold could be shown, the seriousness of Mr. K.'s disease led me to decide for detoxification treatment with DMPS. I refrained from conducting dental restoration work, since Mr. K. was not yet ready for this. The treatment and after-care brought temporary relief for as long as nine months, until both tremor and depression returned worse than before.

Mr. K. approved to have dental restoration work done, which proceeded with all necessary preventive measures in place. After this, he underwent detoxification with DMPS under the administration of supplemental zinc, and vitamins C and E. As early as four weeks later, his tremor decreased and his mood improved—Mr. K. gained hope. Treatment continued for nine months. At first, he got DMPS injections once a week. Later this was reduced to once every fourteen days and finally to only once a month.

About one year later, his wife called me on the phone to tell me how well her husband is doing now. I have, however, recommended regular LTT checkups to keep an eye on the metal levels in his blood. One happy side effect of this all was that it saved their marriage.

Since it is a rather exceptional case, I would like to further elucidate Kurt K.'s case. On the one hand, it shows that norm values (reference values) are essential. On the other hand, however, there are people hypersensitive to exposure to even the most minute traces of foreign substances at levels far below the norm values. In his book "The Boundlessness of Threshold Values", A. Kortenkamp discusses the significance of reference values (67.).

The art of medical healing is to discover each patient's individual features in order to decide on a correct course of treatment.

Despite the outstanding progress made by medical science over the past one hundred years, each diseased patient must be examined and treated individually—this has been known to shamans, healers, and medicine men for several thousands of years.

What lessons can we learn from this case? Our body does not know such a concept as "dubious hypersensitivity". Exposure to heavy metals does not only cause health damage by injuring cells. Even the smallest traces can trigger a hypersensitivity reaction, leading to a weakened immune system, which then goes on to cause disease.

Those are the results that give me a certain serenity and help me to bear the still widely negative stance taken by many conventional physicians. These experiences also continued to encourage me to accept new requests for expert testimony, in full knowledge of the fact that the demands for carrying the costs of treatment will be denied by the health insurance companies, and in spite of the mostly poorly justified reports prepared by the MDK.

7.2.6. Malignant brain tumor (glioblastoma)

Anna S., 62 years old

Ms. S., who had always been in love with her life and looked much younger than her age, suffered from an aggressively growing malignant brain tumor. Within a period of four weeks she was no longer able to walk properly, could not climb the stairs, and constantly suffered from dizziness. It was only after these four weeks, in December 2007, that she first consulted a doctor. CT and MRT scans of the skull were made, which turned up the shattering diagnosis of a tumor in the right cerebral hemisphere.

Ms. S. immediately underwent surgery in a university hospital's neurosurgical department. The pathological-histologic diagnosis was glioblastoma, which is one of the most rapidly growing reticulated brain tumors. For this reason it is often not possible to radically remove these types of tumors. Regrettably this was also the case for Ms. S. A second surgery was again unable to completely clean away the tumor—it had been growing more rapidly again and penetrated the brain even more deeply than was visible on an MRT scan. Post-surgery, she developed paralysis of her legs and problems with voiding her urine bladder and bowels.

In light of the dramatic course taken by her disease, she agreed to my suggestion to test the tumorous tissues for presence of metals. These tests found the carcinogenic metals aluminium and nickel, as well as the potentially carcinogenic tin, mercury, and indium. Genetic testing revealed damage (deletion) the gene encoding for Glutathione-S-Transferase GSTM1, an enzyme necessary for removing metals from the body.

Due to the rapid tumor growth and the severity of the disease, my suggestion to her attending physicians in the university clinic was to try one final option: detoxifying the patient with the help of oncopheresis[19]. My intention was to detoxify the tumor and the patient's whole body to stop the tumor from growing even further and at the same time take away its source of nutrition.

My own studies (126.) in the context of my Ph. D. research had revealed that one can influence the speed of cell growth in dependence of which nutrients are supplied to the cells. It is even possible to fully halt cell growth if one deprives them of substances essential for their proliferation. This procedure leads to stalling or stopping cell growth through cell division. Every cell clearly needs specific growth factors, which we refer to as promoters of growth. It appears that carcinogenic cells play a calamitous role in this process.

As expected, my suggestion was rejected and Ms. S. died two months after her tumor was discovered, despite two operations, chemotherapy, radiation therapy and ensuing rehabilitation therapy.

All metals are capable of passing the blood-brain barrier and damage the central nervous system. Once again, we found the carcinogenic metals nickel and aluminium in the tumorous tissue.

---

19 Oncopheresis is a new concept, first used by me in an article on the influence of heavy metals as well as genetic and immunological factors on the development of malignant tumors (134.). R. Straube, now head physician in the INUS medical center (a clinic in the "Regenbogen" health park in Cham) also used this term. It is a form of therapeutic apheresis that I recommend for detoxifying patients suffering from malignant tumors in the presence of proven exposure to metals (see pages 149 and 150).

Here I would like to make a passing remark: it is rather remarkable that, after doing scientific work in a university clinic (department of pharmacology), one can find use for this biochemical experience as much as forty years later. All energy I invest in my cases is paid back to me and never gets lost. This realization means a lot to me and helps me in particular when things appear to be going all wrong.

7.2.7. Acute allergic reaction to metals (hypersensitivity) after surgery

Klaus B., 70 years old

In January 2005, a small polyp was found during routine bowel endoscopy, proliferating inside neoplasm. Mr. B., however, had no complaints. He was diagnosed with a tumor in the large intestine (colon carcinoma), which urgently required surgery. Since the case involved a stadium 1 malignant tumor, full recovery was expected.

Surgery proceeded without any complications. About 20 cm of the large bowel were removed. On the third day post-surgery, the area surrounding the wound started reddening, which then spread to all of the surrounding tissue. I alerted the surgeons to this possibly being an allergic reaction to the metal clipses use to close the wound on Mr. B.'s abdominal skin. The wound had been tacked—the usual procedure today. After eight days, Mr. B. was discharged from hospital, despite significant reddening and swelling of the whole area surrounding the surgical wound.

One day after arriving back home, the whole seam on Mr. B.'s abdomen split open and he had to be readmitted to hospital immediately. Mr. B. developed a severe inflammation of the abdominal cavity (peritonitis) with intestinal obstruction (ileus) due to all of the small intestine clogged together as if in a lump. He had to be reoperated.

It turned out the "burst belly" had been caused by the metal clips, which contained amongst other metals nickel and chrome, i.e., toxic and carcinogenic metals. These clips had caused an acute allergic skin reaction that caused the surgical wound to split open. Mr. B. spent a total of six weeks in hospital. Thanks to the rapid action by his attending physicians, all ended well and today he is symptom-free.

As this case demonstrates, metals can not only cause chronic but also acute hypersensitivity (allergies). Hence, my advice is to be very careful with the use of metal clips in surgery. Patients should be tested for acute and chronic hypersensitivity to metals before those clips are used. If this is not possible due to time constraints, I recommend not using them.

Acute allergies are increasingly more common among people with a weakened physical defenses (immune system), which means that this is the case for practically every patient post-surgery. In the case of Mr. B., there existed a known allergy to some metals. However, when recording his medical prehistory, he was not asked after such a type of allergy and he was not informed that his abdominal skin would be closed with metal clips. Mr. B. himself in turn, because he had not counted with having to undergo surgery, had forgotten to show his allergy pass. The question as to the connection between abdominal surgery and allergy to metals was never asked. Such severe consequences as in the case of Mr. B., would certainly not have occurred if the attending physicians had taken more time for his case and been more diligent in informing him.

7.2.8. Six breast operations due to mammary carcinoma

Ulrike G., 80 years old
I knew Ms. G. for many years. However, I only had treated her in my capacity as environmental physician for two years (since 2008). From her prior history, it is worth mentioning that she had had two miscarriages, after which she had abandoned her wish for children. She had a happy marriage and never was seriously ill. Eleven years before the diagnosis cancer was made, she underwent thyroid resection to remove an benign tumor (adenoma) in her thyroid.

First operation
In February 2002, she was diagnosed with a mammary carcinoma in her left breast. Her tumor was radically removed during surgery, which also—as it is said in the jargon—cleaned out the lymphatic nodes in her armpit. The interdisciplinary tumor conference decided to follow-up with hormone therapy and supportive radiation treatment. In cases of complicated tumors, not just one or two doctor but rather a so-called tumor conference, with

several medical specialists from different fields, make the medical decisions. These tumor conferences are a very good institution.

Second operation

One year later, the patient was diagnosed with a mammary carcinoma in her right breast that also affected the lymphatic nodes. Once again, the tumor was surgically removed and the lymphatic nodes in her armpit were cleaned out. Treatment proceeded with hormone therapy and radiation treatment for her right breast.

Third operation

In July 2008, Ms. G. developed red spots on her left breast that rapidly increased in size. This time, the pathologic/histological diagnosis was angiosarcoma. This is a malignant skin tumor that can develop after surgery. In this case, the patient's left breast and the fascia of her left pectoralis major were removed. The wound was closed with the help of a skin graft. Ms. B. did not have to undergo radiation treatment.

Fourth operation

In October 2008, red spots once again developed, in the vicinity of the seam on her left breast. The updated diagnosis was: secondary tumor to an angiosarcoma, developed after radiation treatment. The surgery protocol described it as first local recurrence of an angiosarcoma. Once again, the tumor was radically removed. No metastases were found. Due to the patient's advanced age, the tumor conference decided against radiation treatment.

Fifth operation

In January 2009, a second local recurrence to the angiosarcoma developed in the presence of numerous scattered foci. Once again, the tumor was removed surgically and the defect was closed with a skin graft. Once more, no radiation treatment was given. Should the tumor progress, the patient should be given Avastin mono (a chemotherapy drug to treat mammary carcinomas).

It was during this phase of her treatment that I entered the case, on the request of the patient. After consulting the attending physicians, I ordered her empty-stomach urine tested for presence of heavy metals, as well as the transplanted skin tissue and the tumorous tissue. Her empty-stomach urinalysis found exposure to nickel at levels of 3.4 µg/l. The reference value for nickel is > 2.5 µg/l. Levels of the other metals were all within acceptable norms. In the healthy skin graft tissue that surrounded the tumor, the following elevated levels were found: copper 180 µg/kg (5–50 µg/l)[20] and zinc 3,290 µg/kg (> 140 mg/l). In the tumorous tissue, we found the following pathological values: copper with 170 µg/kg, zinc with 3,530 µg/kg, and tin with 160 µg/kg (< 2 µg/l).

Approximately two months later, since the case concerned a local relapse with partially marginal resection edges, and because no other treatment options presented themselves, renewed radiation treatment was considered an option, despite the patient's advanced age. The case was discussed at a tumor conference, where my objection that the angiosarcoma was caused precisely by the previous radiation treatment was not taken into consideration. The conference argued that a new piece of radiation equipment would be used that should cause less side effects. In addition, the extreme malignancy of the tumor mandated renewed radiation therapy.

As usual, treatment was spread over thirty sessions over a period of six weeks. Despite proper skin care and all other precautions taken, the treatment caused a complete burning of the patient's skin and caused her terrible pain all through the treatment.

But that was not the end: in June 2009, to the patient's and her husband's dismay, red spots once more appeared in the area of her left breast. I myself was distraught, even though my previous concerns had prepared me for this already third post-radiation-treatment local relapse of an angiosarcoma within one year. It was something that I had never seen before. It clearly was an extremely malignant and rapidly developing tumor.

Sixth operation
This time, due to the carcinoma's large extent, surgery was conducted in a clinic that specializes in plastic surgery and treatment of highest-degree burns. Chest wall resection was performed with partial removal of the fourth rib on the left, in conjunction with a so-called split skin graft with the skin taken from the left upper thigh.

---

20 Reference values = normal values. No reference values for metal levels in tissue exist. Hence, I compared the values found to the reference values in empty-stomach urine to have at least some indicators for interpreting the results of the tests.

Pathological and histological examinations confirmed the diagnosis of an extended angiosarcoma. Once more, I ordered metal exposure tests on the tumor-free tissue. In the tumorous tissue, the values for three carcinogenic metals—aluminium, lead, and nickel—were significantly elevated compared to the norm values measured in the urine. In addition, exposure to the potentially carcinogenic metals tin and silver was found.

In the healthy tissue only the aluminium and nickel levels were found elevated, with lead being within accepted norms. Exposure to tin was found to be identical in both types of tissue. Silver, with 90 µg/kg vs. 170 µg/kg, was only half as large in the tumorous tissue. The following months turned out to be a *via dolorosa* for Ms. G. and all concerned, to an extent that I have never seen in my more than forty-year-long medical career.

Ms. G. suffered intolerable pain. The tumor effectively spread over the whole of her left breast area and eroded her skin tissue as well as the newly implanted skin graft. Since the patient was fully conscious throughout the whole ordeal, it was undesirable to knock her out with morphine. The medical team searched for a middle road, putting the patient into twilight sleep. However, this did not always succeed, since she needed relatively large amounts of morphine and other analgesics in order to keep the pain under control.

Six weeks post-op, Ms. G. was discharged from the specialized hospital and transported to a clinic in her hometown, where she was lovingly and professionally taken care of by all of the nursing staff. Due to her pain, the dosage of morphine had to be continuously increased. In spite of all, the skin tumor continued growing. Eventually, she was put into twilight sleep and finally into a full medical coma. After six weeks, she was transported to a nearby hospice, where she died one week later.

In this context, two points are worth mentioning:

1. Although the angiosarcoma appeared in conjunction with the initial radiation treatment that was conducted in 2008, the patient underwent renewed postoperative radiation therapy after a second local relapse. With hindsight one can ask whether this was the right decision.
2. Laboratory tests on the patient's empty-stomach urine showed exposure to nickel.

If exposure to nickel is present, patients should not be subjected to radiation treatment in the first place, because this can lead to severe burns, which can go on to lead to recurrence of the tumor. I explained this fact to the attending radiologist for the second course of radiation treatment. Apparently this was news to both him and his colleagues in the tumor conference. Hence, this case shows once more how little attention is given to an environmental-medical point of view within contemporary conventional medicine.

This example, and many similar cases, were very taxing to me. At the same time, they provided motivation to engage myself particularly intensively for people with environmental damage. Within our society today, these patients usually are the weakest parties, who cannot defend themselves. They have no lobby, no money for drugs needed for detoxification treatment, and no money to pay for dental restoration work. Almost all of them sooner or later lost their jobs due to their diseases, often when still at a young age.

Because these examples are in no way individual cases, I conducted numerous studies to answer the question as to whether harmful substances that come from the environment are responsible for our so-called diseases of affluence and for cancer? In chapter eight, I would like to report several of those studies, and discuss their results. Each case includes critical assessment of the results.

I am aware of the fact that this book walks the thin line between personal experience and sober scientific research together with its results. I ask for your understanding when emotions often dominate. Without passion, belief, and conviction of being on the right path, it would not be possible to keep up this kind of work for more than 25 years.

The individual doesn't feel they have a chance to defend themselves during the whole procedure. There are, however, both ways and means. My hope is that this book will help to show them.

8. Eight environmental-medical studies from daily practice

8.1. Evaluating a questionnaire from KV Dortmund

8.1.1. Introduction

Is it possible for a patient to recognize the causes of their environmental disease by themselves?

Despite modern analytical methods, the discussion about the question of the health damage caused by heavy metals and other chemical noxa proceeds under a lot of controversy. It hardly plays any role in the practices of private general physicians nor even in hospitals and university clinics.

This is very surprising. The examples that I present, and similar ones from every individual practice, force us to conclude obvious connections between exposure to harmful substances and diseases. This should actually be apparent to many of my colleagues, in particular since the media continuously report those cases, as exemplified by the school buildings with PCB emissions.

Each year, several hundreds of thousands of chemical substances are estimated to be introduced to the market. Only a trace amount of them are tested for toxic properties. Seals of approval are handed out, even though no sufficient testing on human tolerability has been conducted. The same goes for medical drugs and other medical products, which are not sufficiently tested for biocompatibility (tolerability by humans) before being introduced to the market.

> *Hence, we ask the question: to what extent are the people concerned, the citizens, informed about these harmful substances that we are exposed to in our environment.*

The goal of my first study was to find out if patients with environmental damage can recognize the causes and sources of their disease with the help of a questionnaire. Here, it was helpful that the KV Dortmund had also released a questionnaire that dealt with precisely this question. I used it in my own first study.

8.1.2. Method

A total of 600 patients filled out a written questionnaire. Each of them had a long medical history (up to 20 years and longer), which included numerous visits to doctors and many hospital stays. They consulted me in my practice to find an environmental-medical explanation for their problems and to determine an effective course of treatment. The majority of my patients came from the Märkischer Kreis regions and from the Ruhrgebiet area, but there were patients from all over Germany. All of them had dental restoration material in their mouths and had worn it for years.

From the answers in the questionnaires, three groups emerged, with the following indicated reason for their health damage:

- first group: dental restoration material + other harmful substances from the environment + other reasons
- second group: dental restoration material + other harmful substances from the environment
- third group: dental restoration material

Each patient was allowed to take home the questionnaire before any examination was done, so that they could fill it out in peace. We helped out if any difficulties appeared.

8.1.3. Results

All 600 patients in group one indicated dental restoration material, noxa from the environment, and other reasons as cause of their complaints.

First group

Of the 600 patients in the first group, 61 % complained about fatigue and loss of motivation and 60 % about inner turmoil. 53 % indicated general loss of performance and 45 % suffered from sleep disorders. Between all 600 of them, there were 1,317 psychiatric symptoms, which corresponds to 220 %. 45 % of patients complained about headaches and 34 % about problems concentrating. Vertigo was indicated by 32 % of patients, and 28 % mentioned strange nervous sensations. Neurological problems were mentioned 830 times (138 %). Hence, each patient suffered from at least two psychiatric and one to two neurological symptoms.

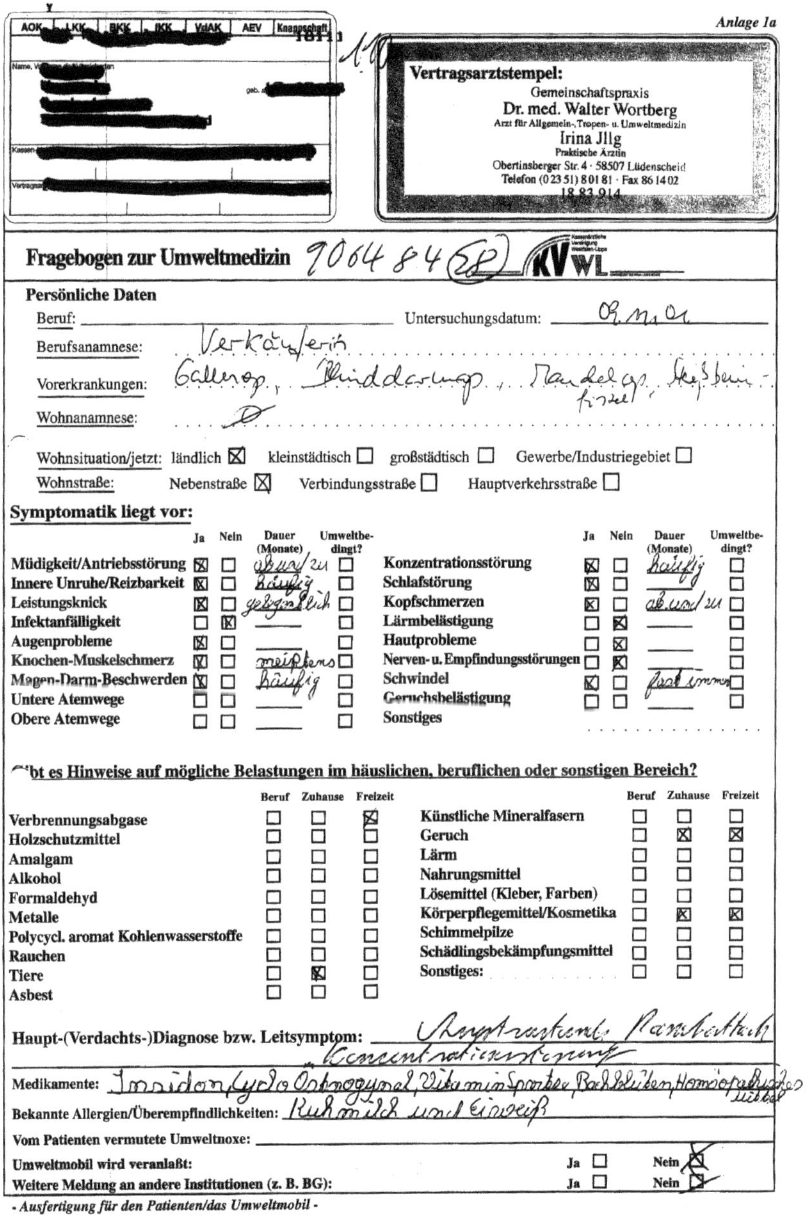

Figure 19: environmental-medical questionnaire

An additional 71 % of interviewees complained about symptoms relating to their sense organs—this is more than two third of the patients. Diseases of the skeletal system and connective tissues, as well as problems with skin and the respiratory system were mentioned in 50 % and gastrointestinal diseases in 43 % of cases.

Second group.
A total amount of 67 % of interviewees reported that other environmental noxa as another possible cause for their symptoms, aside from dental restoration material. Once again, psychiatric complaints and neurological symptoms were by far the most frequent, with 210 % and 62 %, respectively. The spread of psychiatric symptoms was as follows: fatigue 62 %, inner turmoil 61 %, sleeping problems 50 %, and general loss of performance 36 %.

As to neurological symptoms, headaches were most prevalent, with 52 %, followed by concentration problems at 59 %, vertigo with 39 %, and nerve irritations with 22 %.

98 % of interviewees complained about problems with their sensory organs. Respiratory diseases, problems with the skeletal system and the connective tissues, gastrointestinal problems, and skin trouble were again indicated by half of the patients.

Third group
One third of interviewees believed their symptoms to be caused by dental restoration material. Only amalgam was mentioned as used material. To my surprise, other precious metals and/or gold alloys, by now quite frequently used as dental restoration material, was mentioned by hardly anyone.

Once again, psychiatric (151 %) and neurologic (91 %) complaints were the most common ones.

If we compare the frequency of symptoms over the three groups, it becomes clear that each of the symptoms is most common in the first group, closely followed by the group of patients with dental restoration material plus exposure to chemical harmful substances. In the group of people who only wear dental restoration material, the lowest number of symptoms was indicated.

In one third of the cases, amalgam is most frequently indicated as a source. It is followed by smoking/nicotine at 24 %, domestic animals at 18 %, noise and olfactory pollution and cosmetics at 14 %, and fungi at 13 %. Harmful chemicals of any kind were mentioned by 55 %.

Cause of their disease as indicated by interviewees, in %

Amalgam	33
Nicotine	24
Domestic animals	18
Odours, cosmetics	14
Fungi	13
Solvents	11
Combustion gases	9
Nutrition	8
Wood preservatives	8
Formaldehyde	6
Chemical harmful substances	55

8.1.4. Critical assessment

The results of this study are revealing. They show that patients are perfectly capable of recognizing the cause of their disease and trace it back to its source, as long as one asks them the right questions. For a large fraction even the long distance that they had to travel did not stop them from visiting my environmental practice.

The psychological stress experienced by these people must have been very bad—there clearly are not enough environmental physicians. According to the federal government (status of 2007), the fraction of environmental physicians is 1.3 %. For private physicians, this number is far below 1 %.

In overview, it appears that, in the opinion of the interviewed patients, health damage caused by dental restoration material and other environmental noxa is extensive. All organs are affected, but the psyche and nervous systems are under particular attack. We may credit M. Daunderer for being the first to point out these facts in his numerous publications on the topic.

The environmental-medical nature of chronic diseases explains the low success rate of treatment by psychiatrists, psychosomatic specialists, and neurologists. All patients had consulted colleagues from many different specialisms before visiting my environmental practice.

It is clear that dental restoration materials and chemicals play a much larger role than medical specialists and responsible parties within our healthcare system assume. The clinical picture cannot be clarified by conventional medical examinations. The question as to which ingredients in amalgam or other restoration materials (e.g., mercury, copper, tin, silver, gold, palladium, chrome, cobalt, titanium, etc.) are responsible for these symptoms remains unanswered. It becomes clear from our survey that the interaction of dental restoration materials with other environmental noxa leads to a deterioration of the symptoms.

With certainty, amalgam is not the only cause. Instead of calling it an alloy, we should refer to it as a solid fluid, since mercury evaporates. This may explain why many scientific studies that were limited to just the toxicity of mercury as a component of amalgam, were doomed from the beginning. Even mercury itself is harmless, when tested over a short period of time. It should be tested for a period of several years, but who is going to do this?

It is also worth mentioning that noise pollution in the workplace and cosmetics, in that order, were mentioned as causing diseases. Odour nuisances at work, due to new furniture, domestic paints, and due to exhaust gases were also indicated.

In Germany, the danger posed by fungi proliferating in humid apartments and in the workplace has also not yet been overcome. In contrast, many people are aware of the health damage caused by solvents, harmful chemicals, and combustion gases. It surprised me to find out that as many as 10 % of interviewees mentioned nutrition as source of their disease.

> *All in all, there is a big need for action! We need further laboratory test and clinical studies to examine the toxic properties of individual heavy metals and harmful chemicals.*

Since dental restoration materials were the most frequently mentioned source, and hence the cause, of people's complaints, I conducted a material analysis study, which I will describe in the following section. My goal was to answer the question as to what danger is posed by these restoration materials.

8.2. Dangers posed by dental restoration material

8.2.1. Introduction

Material analysis of dental restoration material used in 327 patients from an environmental-medical practice

The amalgam discussion, which I mentioned in section 6.8.2., seems to have influenced our thinking and action to the extent that the dangers posed by other, newer, dental restoration materials are not recognized at all.

Since many manufacturers were no longer sure if amalgam really was harmless, they rapidly developed new materials, always convinced that these were less harmful to our health. In the meantime, the majority of patients with environmental health problems not only wear amalgam, but also have numerous other alloys as well as metal prostheses inside of their teeth. In my clinical study, I tried to answer the question as to what damage these dental materials and their constituent metals pose to our health.

8.2.2. Method

The study used data data gathered from 1994 to 2002. For 327 patients displaying environmental-medical damage, material analysis of the used dental restoration material was conducted. Hence, this number is the basis of this study's data analysis. Without exception, the cases concern patients whose teeth either contained amalgam or other dental material or at least contained it until shortly before the study commenced. The patients' ages ranged from 4 to 75 years, with an average of 43 years. All of them had an extended medical history, ranging from three months up to twenty years.

Comprehensive anamnesis was taken for all patients. Furthermore, each patient was asked to fill out a questionnaire prepared by the KV Dortmund, as well as a letter listing current and previous dental restoration material.

In several exemplary cases, diseases in this group of patients were shown. Only when the composition, and hence the corrosion stability of the used dental restoration material is known, the substances found in urine can be examined for exposure to harmful materials, and conclusions about health damage may be drawn. It is particularly indispensable to conduct examinations for corrosion stability, since recent studies conducted by D. Brune (14.) showed that in the case of amalgam, chemical degradation (corrosion) begins in as little as 30 days.

8.2.3. Results

When the study was conducted, 316 of the 327 participating patients still wore amalgam fillings. The remaining eleven had worn amalgam at some time in the past. Nine of them did not have any of their own teeth and wore partial or full prostheses. One patient was fitted with a titanium implant, her few remaining teeth carrying cement fillings. One other patient only had six carious teeth left, two of them filled with cement.

Numerous participants had fillings made from alloys other than amalgam. Many of them suffered from twenty or more diseases—one patient with actually 42 diagnosed diseases. In total, the 327 patients suffered from 1,700 organ disorders, corresponding to five diseases per patient.

Of the participants, 70 % had six or more amalgam fillings and 17 % had both amalgam and gold alloys, partially consisting of gold/platinum alloys or alloys with lower-cost metals, typically gold and 30 % to 70 % palladium.

The so-called mixed group consisted of 68 patients (21 %) wearing alloys of amalgam plus gold-containing fillings and/or fillings made with precious metals other than gold, or—in some cases—synthetics and/or ceramics. The aforementioned precious metals typically contained palladium (78 %) and other metals such as copper, tin, or indium.

Soft fillings, made out of synthetics, cement, and/or ceramics were fitted in 30 % of the participants, sometimes in combination with heavy metals.

Several patients wore braces on their lower jaw made from precious metals and/or model casts in their upper jaw. Typically, these prostheses consisted of precious metal alloys (for example Biosil, Wironium, Heraenium) containing cobalt, molybdenum, and chrome. With 8 %, 27 patients wore amalgam or a palladium-containing alloy. Six patients

(2 %) had amalgam fillings combined with synthetics and/or ceramics, cement, or porcelain.

One patient worth a specific mention, a 52-year-old female, had restorative material fitted in 24 of her teeth, in part in the form of gold crowns, in part treated with amalgam or cement. She wore a model cast prosthesis made from Biosil in her upper jaw, and her lower jaw was fitted with braces, also made from Biosil. Her teeth had been fixated with a support made from synthetics and four teeth in her upper jaw had been fitted with a crown made from gold-ceramics.

Material analysis of the used dental restoration materials found the following 16 metals: mercury, copper, tin, silver, gold, platinum, indium, gallium, molybdenum, chrome, cobalt, palladium, beryllium, thallium, zinc, and traces of nickel. The patient suffered from 10 different organ diseases, with unknown causes, most notably fibromyalgia, psoriasis arthropathica, and Hashimoto's thyroiditis (a type of thyroid inflammation). Laboratory results showed pathologic thyroid parameters and a slight left-shift in the blood count. Blood sedimentation values were elevated to 20/45.

The photos in figures 21 to 25 show another glaring case, which involves a 65 year old female patient before and after detoxification treatment. Figure 27 and 28 show the dental restoration material that was used. In figure 27, on the bottom right, we recognize a model casting prosthesis for the upper jaw (1.) and on the left top we can see a lower jaw prosthesis in the shape of a clasp (2.). Both of them are on a mount made out of synthetic material. Five teeth have palladium-containing gold crowns (3.). The other teeth are mounted with the help of female (4.) and male (5.) matrixes.

The female matrix is a tight clasp that looks like a claw. In this case, it is made from a precious metal alloy called Ceramicor. According to its manufacturer (Cendres+Métaux from Switzerland), it contains 60 % gold, 13 % platinum, 20 % palladium, 1 % iridium, and 6 % other materials. The male matrix, which is the actual support, is called Doral and is made out of 15 % gold, 49 % silver, 22 % palladium, and 14 % copper.

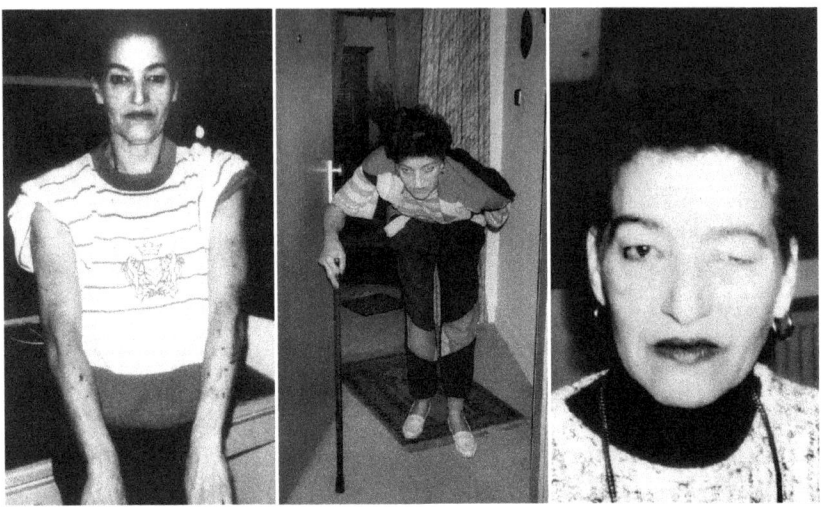

Figures 20–22: 65-year-old female patient with neurodermitis (left), fibromyalgia (middle) and Graves' ophthalmopathy* (right) before treatment

*) Graves' ophthalmopathy is an autoimmune disease of the eye muscles and the orbital connective tissue that develops in the presence of Basedow's disease, i.e., it is connected with hyperthyroidism.

Figure 23 (left): toothbrush after single use, before and after treatment
Figures 24–25: the same patient after treatment

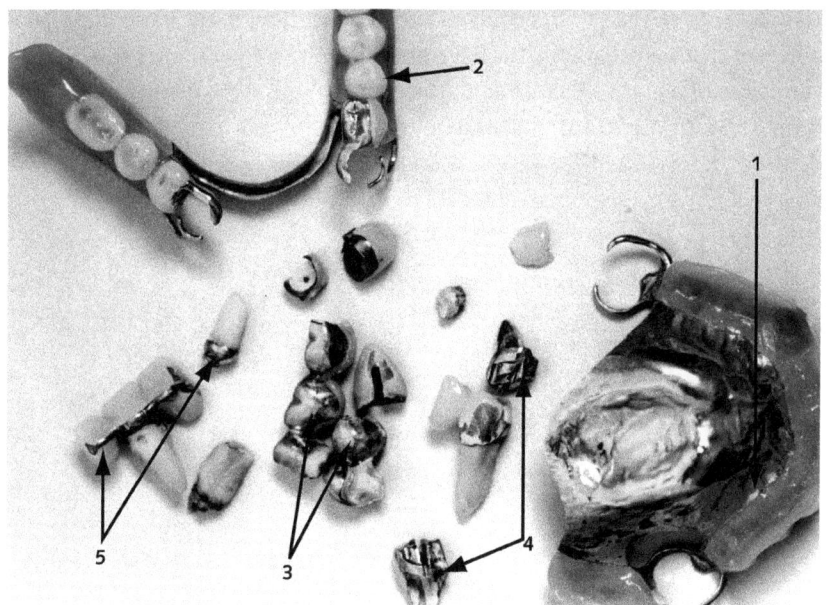

Figure 26: dental restoration material, made out of Deva 4[21], from the mouth of a 65-year-old female patient.

1. model casting prosthesis for the upper jaw, made out of Deva 4
2. lower jaw prosthesis, made out of Deva 4
3. five teeth with palladium-containing gold
4. two teeth with female matrixes
5. four teeth with male matrixes

This 65-year-old patient had 29 diseases (e.g., neurodermitis, fibromyalgia, and Graves' ophthalmopathy) affecting in total 14 organs. She had been wearing amalgam fillings for more than 30 years. In 1989 these fillings were removed after suspected exposure to heavy metals. They were replaced with crowns made out of other precious metal alloys, in particular gold ceramics.

Due to permanent inflammations inside her oral cavity (inflamed gums, suppuration of the dental roots) many teeth, including the restoration mate-

---

21 Deva 4 is a palladium/gold alloy composed of 51.1 % gold, 38.5 % palladium, 9 % indium, 0.2 % iridium, and 1.2 % gallium.

rial and crowns, had to be removed later. As mentioned, the patient received a model casting prosthesis for her upper jaw and a lower jaw prosthesis in the shape of a clasp. According to the dentist, both were made out of Deva 4 on a mount made out of synthetic material.

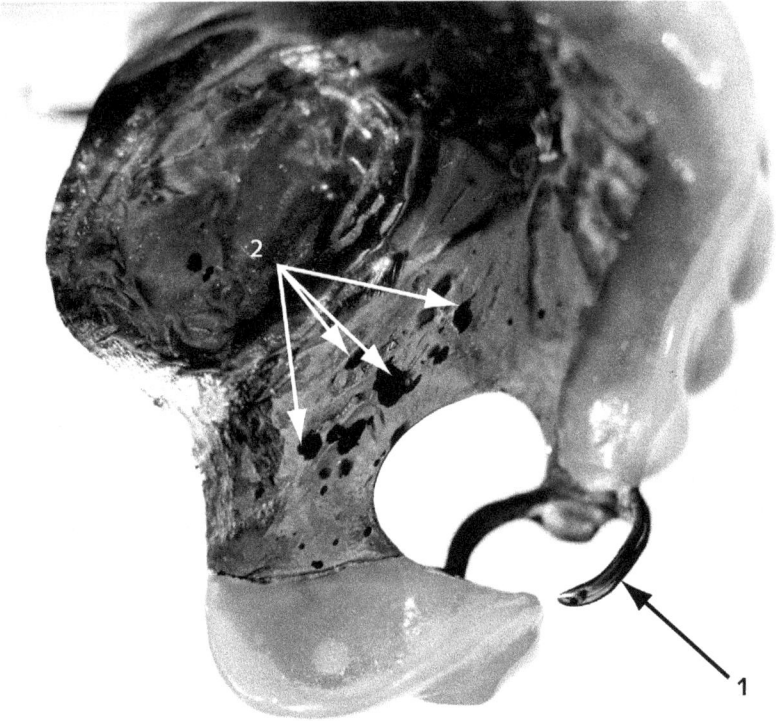

Figure 27: detail photo of the model casting prosthesis made out of Deva, on a mount made out of synthetic material, after having been worn by the patient for nine years.
1. *clasp, without corrosion marks*
2. *upper dental plate, with numerous holes (corrosion marks = rust stains).*

Laboratory tests only revealed elevated mercury and copper levels in the blood and exposure to mercury and copper in the empty-stomach urine. In addition, the patient was allergic to the heavy metals gallium and nickel. Trace amounts of nickel (less than 0.1 %) are present in almost every alloy and hence do not need to be listed as a constituent.

The patient first underwent surgery in 1955 to remove a benign nodule in her thyroid (goiter). In 1995, she had to be reoperated to treat hyperthyroidism (Graves' disease). The surgeons removed her thyroid, after which she developed immunodeficiency of the thyroid (autoimmune thyroiditis) for which she underwent two courses of radiation treatment.

Figure 28 shows the model casting prosthesis worn by the patient, photographed in more detail. Even though the lighting shows the Deva-made surfaces on the upper dental plate and the two clasps, which usually almost sparkle like silver, seem partly light brownish and partly reddish-brown, it is still possible to see individual details. Near the clasp (1.) the gold-palladium alloy and its coating are fully preserved. To the right lower area, however, we can see dark-brown, rust-like stains some of which are spotlike, whereas others are more of an extended area (2.).

The dental expert witness describes these spots as follows: "The surface of the model casting prosthesis is, contrary to expectation, dull and shows cavities that are visible to the eye." These cavities are abrasive and corrosive and can hence be called rust stains. (An expert witness would never use the word "rust".)

The abrasions are even more noticeable in Figure 26, which shows two of the patient's toothbrushes. The toothbrush to the right is unused, whereas the one on the left was photographed after single use. It shows a grey-blackish coating that was rubbed off the teeth. According to the expert witness, this probably is silver, an important ingredient of amalgam.

Further toxicologic tests (multi-element analysis) on the tissue of the lower jaw found exposure to the following eight heavy metals: copper, silver, palladium, platinum, gold, cobalt, tin, and gallium. Particularly noteworthy are the high levels of copper at 5,520 µg/kg (normal levels are 5 µg-50 µg/l/kg) and gold at 1,060 µg/kg (normal levels < 0.2 µg/l).

If we analyze the various alloys used in this patient (amalgam, Deva 4, palladium/gold crowns, Ceramicor, and Doral) we count as many as 12 heavy metals and elements.

The upper and lower jaw are accumulating organs, which are often overlooked after dental restoration work, or are left uncared for out of ignorance. As concerns detoxification therapy, this accumulation slows down removal (elimination) of the metals from the bones, necessitating further periodic detoxification regimens.

8.2.4. Critical assessment

The majority of patients in our study was not fitted with one single type of dental restoration material, but rather had a smorgasbord of heavy metals and precious metals, as well as nonmetallic dental material inside their mouths—this is remarkable. Only 38 % of patients wore amalgam as single material, whereas 62 % wore amalgam plus another filling material.

Worth pointing out is the large fraction of amalgam-wearing patients with more than six fillings. A number as large as this is admitted even by proponents of amalgam to carry a higher risk to the health of individual patients. Additionally worth mentioning is the added number of patients wearing alloys made out of gold or other precious metals.

This large amount of foreign substances that were found in patients' mouths and the large number of diseases in these people—as testified by the examples of two severely ill patients with 16 and 12 metal- and plastics-containing dental prostheses—strongly urges us to suspect a causal connection between the diseases and the used dental restoration material.

The following facts further corroborate this suspicion:

1. In both cases, exposure to heavy metals was determined in urine.
2. In the case of the 65-year-old female patient, a type i. v. chronic allergy to gallium and nickel—metals used in her dental restoration material—was found.
3. Measurements on tissue samples taken from the patient's lower jaw demonstrated high exposure to copper (5,520 µg/kg compared to a normal value in urine of 5–50 mg/kg) and gold (1,060 mg/kg versus 0.2 µg/kg). As mentioned before, no reference values have been established for heavy metals in tissue, but it can be safely said that amounts of gold and copper as large as were found in these tissue samples and bodily fluids are beyond all acceptable norms.
4. The metals found are or were part of dental fillings that the patient wore either at the time of the examination, or in the past. Hence, the measured exposure may be caused by the used materials.

Despite the high quality of many alloys in use, electrochemical corrosion processes had caused heavy metal ions to dissolve over time, which were then transported via the respiratory system, blood, and lymphatic system into the patient's organs. (82., 83., 122.)

We will not further discuss the concepts of electrochemical corrosion and electrolysis. However, it is known even to laymen that there are small voltage differences between different metals which, in the presence of a so-called electrolyte (in this case the patient's saliva), cause metal ions to dissolve from their matrix.

Numerous dental *in vitro*[22] and, to some extent, *in vivo* studies have demonstrated the described mechanism for many dental alloys in regular use. (14., 82., 83.). C. Mortimer describes in detail how these electrolytic and corrosive processes develop. Hence, corrosion and electrolysis are not only relevant to industrial processes with respect to the durability of materials but most certainly also play a crucial role in wear and biocompatibility of dental restoration material and hence in the human body's exposure to harmful materials.

Processes such as mechanical abrasion due to chewing, nightly bruxism, brushing teeth, and so-called material fatigue are further mechanisms that influence decomposition of dental fillings, in particular once the used material ages beyond its expiry date.

To this we can add corrosive attacks on the materials by nutrition and saliva, which effect is to some extent increased by bacterial infections and elevated temperature inside the oral cavity during any periods of illness that are accompanied by fever, all of which also damage the fillings. Hardly any surface is safe from corrosive processes and we can assume that this applies to all dental materials in use (metal alloys, but also plastic materials, such as synthetics, ceramics, and cement).

Both the manufacturers and the public health department provide guarantees and written information according to which every single material is tested for tolerability, and each of them is certified with a seal of approval, the CE and/or TÜV seal. However, these tests will not find the problems mentioned above and do not point out the relevant risks clearly enough.

---

22 In vitro (from the Latin: in glass) designates organic processes that take place outside of a living organism—this in contrast to processes that occur in a living body. The natural sciences use *in vitro* experiments that can be conducted in an artificial environment outside of the living organism, e.g., in a test tube.

Instead, they lull both patients and the whole dental profession into a false sense of security.

Testing procedures for biological tolerability (biocompatibility) are dictated by DIN-Norm 1 3930. Therefore, manufacturers and importers are not required to conduct the relevant tests, since this is regulated by the federal law on medical drugs. In my opinion, this is a bad solution. Dental materials bear no relation to medical drugs, but they are both substances foreign to the human body. For this reason, one must assume that they are not tolerated and can cause health damage. It follows that, if these substances are not tested for biocompatibility prior to their use, side effects in the shape of minor and serious health damage is effectively preprogrammed.

For all biomaterials, multistage selection and testing procedures are in place:

- Stage 1: *in vitro* screening (e.g., cell culture testing as described).
- Stage 2: *in vivo* tests (e.g., implants) with test animals.
- Stage 3: clinical tests on patients, in a controlled environment, for a limited period of time.
- Stage 4: widespread clinical application of the material, following approval by the legislative councils, and after epidemiological studies have concluded with positive results (successful application).

The more primitive the biological testing systems are, the less clear the relevance of the testing procedures and the extension of their conclusion to humans, becomes. As of today, there are no international binding or standardized procedures or norms for biocompatibility testing.

> *Can it be that the laxity displayed by testing committees and the parties responsible for the federal laws on medical drugs is the reason why no progress has been made in this area?*

One should re-read these sentences a few times and wonder what this implies for our health. When it comes down to it, the life of many thousands, even many millions of people depends on the quality of the aforementioned testing procedures—and there still are no standardized methods. In my opinion, this counts as culpable neglect by the responsible politicians and doctors, and the federal government, whose duty it is to supervise these procedures. In light of the fact that almost every single adult in Germany wears metallic materials in their teeth, one must ask how to defuse this time bomb.

So, what does stage 1 screening involve? Tests are conducted over a period of three to seven days, during which the materials are suspended in a buffered medium, e.g., a 0.1 molar lactic acid/saline solution (pH = 2.3), or in fetal bovine serum, at a temperature of 37°C. The measured quantity is ionic release from the materials, in µg/cm^3. As an example, we mention the use of mouse fibroblasts[23].

Keeping in mind that dental fillings typically spend up to ten years (or longer) being subjected to strong mechanical stresses, rather than three to seven days of testing, biocompatibility certification appears rather dubious. In addition, most patients not only have one single metal alloy filling but are typically fitted with several dental prostheses, containing different metals that have quite distinct electromotive series. The study also mentions this fact, and once again these findings are not taken into account during any of the four stages of testing.

My own research has shown that over the past years not only the number of amalgam fillings per patient has increased, but also the amount of non-amalgam-containing materials. Hence, our issue is not just limited to amalgam but can be classified as a general problem with dental materials.

From this all we conclude that, due to the short duration and the used testing substrates, the described corrosivity and biocompatibility tests that manufacturers conduct in order to qualify for a TUV certificate, as well as the monitoring procedures used by the health and drug authorities, are simply not sufficient as testing methods. Even tests running for more than 30 to 60 days do not reflect contemporary *in vivo* conditions. Also, patient groups at particular risk (children, pregnant women, sick people, the elderly) are not at all taken into consideration, despite them being at elevated risk.

My study showed five different organ diseases per patient, which begs the question:

> *Does there exist a correlation between dental restoration material and the increase in health damage? The examples of the two patients with the many filling materials illustrate that this question is justified: the number of medical conditions and organ disorders increases with the number of fillings.*

---

23 Fibroblasts are connective tissue cells that play an important role in the synthesis of intercellular material.

This observation is corroborated by measurements of heavy metal levels in urine and/or by chronic heavy metal allergies shown in patients. It is known that the incidence of such chronic allergies against nickel, mercury, gold, silver, palladium, chrome, and other heavy metals is on a continuous increase. Here too, we may ask whether anyone can justify the increasing use of such materials in dental restoration work.

In the preceding discussion I mentioned a few cases of intrauterine foetal damage. The following study deals with the question of prenatal damage to the foetus.

8.3. Intrauterine foetal damage caused by maternal exposure to heavy metals

8.3.1. Introduction

Studies conducted by E. Bonnet (13.) and I. Gerhard *et. al.* (34., 35., 36., 37., 38.) on the connection between mercury exposure and fertility disorders show that, if the mother wears metal fillings during pregnancy, mercury can enter the foetus through the placenta, which can lead to serious health problems in the child. These conclusions have been confirmed by findings from the Munich University Institute of Forensic Medicine, which studied mercury levels in foetal brains. The topic was also studied by Koos (65.), Merzoug (81), and White and Wiebe (123., 124.).

Our case studies show that not only mercury, but also copper, tin, and organic tin compounds are transported from the mother into the foetus, causing serious health damage to neonates, infants, and children. In his book, "Applied Environmental Medicine" (107.), Schulte-Uebbing concludes that: "As a possible catalyst for immunodeficiency and other irreversible damage, amalgam is highly significant as a harmful substance, in particular to the fields of gynaecology and obstetrics."

The study aims at two targets: first, we verify the results obtained by Bonnet and Gerhard within a general and environmental-medical practice. Second, we want to find out whether the mentioned examples of intrauterine transferal of copper and tin (which are also constituents of amalgam), as well as of organic tin compounds, are isolated cases. If this is not so, we must conclude that these compounds too can lead to pre- and postnatal health damage to neonates.

8.3.2. Method

The study comprised data of 57 children of in total 34 mothers, i.e., one to two children per family. The children ranged between 2 and 20 years of age, with an average age of nine years. The mothers' ages ranged between 16 and 60 years, with an average of 39 years. One of the mothers was just 16 years old and can strictly speaking be considered one of the children.

It was set as a precondition that the children had been subjected to a thorough general-medical and medical-specialist examination and had received appropriate treatment, if necessary, prior to their participation in the study. In addition, written proof was required showing that none of the children wore any amalgam in their mouth and that each mother had worn amalgam fillings in the past (or still wore them) or, possibly, underwent dental treatment during her pregnancy.

Empty-stomach urinalysis showed heavy metal exposure in all of the children. At the time of the study, the majority of participating mothers wore amalgam fillings or had them removed during or after pregnancy.

For each child and both of its parents, a thorough anamnesis, comprising family history, social background, and professional history, was taken. In addition, a record of the parents' dental status before and during pregnancy was taken.

The study commenced with extensive clinical and laboratory tests on the children and their parents. Urinalysis for heavy metal exposure was conducted by the Bremen Medical Laboratory, for 57 amalgam-free children. After this, empty-stomach urinalysis was performed. If no exposure to metals was found, DMPS (Dimaval, manufactured by Heyl (102.)) mobilisation tests were conducted (see chapter 9.8.2). DMPS is a chelating agent that chemically binds to various metals after which the formed complex is excreted via the bowels or the kidneys. Hence, Dimaval is a proven metal detoxification agent and is used in cases of acute poisoning and, by now, is also a well-tried treatment for chronic poisoning.

Patients received either an infusion of 3 mg Dimaval per kg of body weight or a capsule containing 10 mg Dimaval per kg body weight. After this, they drank 150 ml water. Two hours later, a second urine sample was taken, of which 20 ml were used for analysis.

Depending on degree of toxic exposure and individual tolerance, the children underwent detoxification treatment with Dimaval and/or DMSA. Once their condition started improving and measured exposure levels had decreased, post-treatment continued with medicinal plants such as coriander herb (Paracilantro drops), Solidago, Hepatika drops and microalgae (Beta-Reu-Rella), as well as supplements containing Vitamin D3 and zinc. After detoxification, urine samples were taken and tested for metal exposure every six to twelve months (see chapter 10 for further explanations).

No multi-element analysis (MEA), i.e., a test for heavy metals used in alloys, was conducted since this was not covered by the children's health insurance. However, these tests were conducted for some of the mothers.

Any measurement larger than its reference value can be considered an indicator of metal exposure. The so-called load factor is calculated as the quotient of the average of the measured values divided by the reference value.

8.3.3. Results

Heavy metal exposure levels in the children

In the children, empty-stomach urinalysis levels showed average load factors of 1.3 for mercury, 1.5 for copper, and 2.6 for tin. After treatment with DMPS the measured values were 2 for mercury, 4 for copper, and 2.5 for tin. Values for both measurements were subject to a significant fluctuation.

The aforementioned fluctuation was found to be even larger for organic tin compounds, with elevated levels of mono- and dimethyltin and of mono- and dibutyltin. In contrast, no organic mercury (methylmercury) could be found, and it had not yet been possible to analyse the samples for levels of organic copper compounds. Once again, the values for mercury were the lowest, at 1.2. The highest values were found for monobutyltin, at 30.

For 67 % of the children, exposure to copper could be shown, exposure to tin for 49 %, and five of the children showed exposure to mercury. Organic tin compounds were measured in 30 % of the participating children, most frequently monobutyltin (23 %). Exposure to methyl mercury (organic mercury) could not be found.

After detoxification, all values returned to within accepted bounds and this was confirmed by follow-up examinations up to three years after the study.

Heavy metal exposure levels in the mothers

For the mothers, load factors for mercury, copper, and tin as measured in empty-stomach urinalysis were lower than for the children, both before and after DMPS treatment. The same result was found for organic tin compounds.

Elevated heavy metal levels were found in all 34 participating mothers. Of them, 40 % had elevated tin levels, 35 % in the case of mercury, and 33 % for copper. In 10 mothers (29 %), elevated organic tin compound levels were measured, and methylmercury was found in 14 % of the cases. A total number of 15 of the 34 women showed exposure to organic tin compounds.

During analysis of the load factors for individual heavy metals, it was noted that 29 mothers (85 %) showed additional exposure to metals such as palladium, platinum, cadmium, and cobalt.

Neurological and psychiatric symptoms

In the group of children, 88 % suffered from neurological and psychiatric conditions. Of the latter, inner turmoil (21 %) and sleep disorders (17 %) were most prevalent, followed by somnolence and fatigue. The most frequently made diagnoses were ADD, hyperactivity, anxiety, and bedwetting.

The most frequent neurological complaints were headaches (33 %) and problems concentrating (25 %), followed by vertigo and developmental disorders. The most frequently encountered diseases were migraine, vertigo of unknown etiology, speech disorders, and mental and physical developmental disorders with unclear causes. A total number of 57 children registered more than 91 psychiatric and neurological symptoms, which means that each child suffered on average two symptoms.

Organ disorders in the children

Our survey shows that all organs are targets of diseases, most frequently (93 %) the nervous system (restless leg syndrome, epilepsy, stuttering, mental and physical impediments) and the psyche (ADHD, 72 %), followed by dermatological disorders (35 %) such as neurodermitis, eczema, ichthyosis, acne, allergic exanthema, and hair loss.

32 % of the children suffered from respiratory diseases and recurrent infections, among them asthma and pseudocroup. Immune system disorders, as well as allergies to house dust, mites, medical drugs, and a wide range of

pollens were seen in 26 % of the children. Of the cardiovascular diseases, Raynaud's disease and angina pectoris were strikingly prevalent, with 12 % of participating children suffering either of the two.

With regard to disorders of the sensory organs, children most frequently complained about vision problems, burning eyes, and olfactory disturbances. Concerning urogenital disorders, we found one case of cryptorchidism. Gastrointestinal diseases were represented by a case of gallstones in an 18-year-old girl.

Comparison between children and their mothers
The type of diseases as well as their frequency were fairly similar between mothers and children. Psychiatric diseases occurred in 72 % of the children and in 100 % of the mothers. Neurological complaints were mentioned by 93 % of the children and, once more, by all of the mothers. Dermatological conditions occurred in 35 % of the children and 41 % of the mothers. The numbers for respiratory diseases were 32 % for the children and, once again, 41 % of the mothers.

8.3.4. Critical assessment

In the children, significant copper and tin load factors were measured noticeably more frequently than for mercury. The high copper values may be explained by the common use of copper amalgam in the 1960's to 1980's, when the mothers were treated, which was also the time span before the birth of the participating children. Due to its high corrosion susceptibility, copper amalgam has been replaced with silver amalgam in the 90s.

The childrens' diseases—many of them classified as symptoms—the cause of which was so far not known are clearly not only caused by inorganic but also by organic metal compounds, a fact that so far has not been given due attention. It is, however, a fact of high medical significance, since organic compounds are up to 100 times more toxic than inorganic ones. Hence, I will discuss them in some more detail in the following.

Organic compounds are up to 100 times more toxic than inorganic compounds.
In the human body, organic compounds are formed through e.g., biomethylation, usually by bacteria and fungi located in the bowel, where the compounds are subsequently stored. Once they are released from the bowel,

these organic metal compounds are transported to all organs in the body. Embryos and foetuses react with particular hypersensitivity to the presence of such metals coming from the mother. The formation of the various organic metal compounds is known as transmethylation and transbutylation.

The results of our study confirm our suspicion that in all likelihood not only mercury, but also copper, tin, and organic heavy metal compounds are transported from the mother via the placenta into the foetus, where they cause the damage described before. These observations are similar to the results found by I. Gerhard (34., 35., 36.) and E. Gleichmann (40.).

> *When the same metal exposure load factors are found in 57 children not wearing any amalgam, as were found in their mothers, and when, additionally, similar diagnoses are made for both groups, this cannot be coincidence.*

Another clear indication that metal poisoning caused the diseases is the fact that the health of 80 % of the children, as well as of their mothers, improved after detoxification. These improvements remained two years after treatment, and in several cases the patients were fully cured. Nevertheless, as confirmed by my own studies, checkups and treatment regimens should be followed for up to three years, or longer, after detoxification.

In 85 % of cases, the mothers showed additional exposure to metals other than copper, mercury, and tin compounds. Examples are palladium, platinum, cadmium, and cobalt all of which are probably also transferred to the child (95.).

If intrauterine metal-related damage to the foetus is confirmed, we should also anticipate genetic damage. We may now ask ourselves whether the damage took place within the uterus, before the pregnancy, or even generations earlier. Obviously, the chances of successful recovery are better in the case of intrauterine damage, compared to (genetically) inherited damage.

From these results we conclude that women who desire to have children and who suffer from diseases of unclear etiology should let themselves be tested for heavy metal exposure before becoming pregnant. It is also recommendable to carry out multi-element analysis for inorganic and organic heavy metals for all of the children that suffer from diseases with unknown causes.

Our results also show an urgent need for research into intrauterine foetal damage and genetic damage. We should find an answer to the question of whether noxa other than metals, e.g., chemical contaminants and fungal toxins, may cause intrauterine foetal damage. Given the fact that these substances can pass into the placenta, we may expect this to be the case.

On analysing the individual examples, together with the three studies discussed previously, we arrive at two questions:

1. What is the role played by tin compounds (organic and inorganic)?
2. Are metals the hidden cause of chronic diseases as well as of both malignant and benign tumors?

In the following section, I will deal with the first question.

8.4. Health damage from tin compounds

8.4.1. Introduction

Excerpt from a lecture that I gave in Kiel on 04.03.2000 on the occasion of the environmental conference and the farewell to Prof. Wasserman.

My lecture discussed health damage caused by organic and inorganic tin compounds, also called organotin compounds. I would now like to sketch how both my specialized research and the lecture came to be.

There are hardly any studies on the health risk posed by organic tin compounds, which are the most toxic metal compounds altogether. One has good reasons to ask why a private doctor specializing in general and environmental medicine decided to tackle such a complex matter.

A certain Dr. med. H., from the medical laboratory in Bremen, was asked to hold a lecture about health damage from tin compounds on the occasion of the retirement of Prof. Wassermann, one of the leading and most famous toxicologists in Germany. As mentioned, he gained fame in particular as the plaintiff's expert witness in the Frankfurt amalgam trial.

Since I appeared to be the only environmental physician in Germany who studies organic tin compounds (organotin compounds), I was asked to report my results. Dr. H. discussed manufacturing processes, after which I presented my report on the health damages caused by these compounds. It was to be my first talk on an environmental health topic.

At the start of my lecture, I pointed out the damage to marine plants (algae) and animals (mussels and slugs) caused by organic tin compounds, in particular tributyltin. In particular, developmental disorders, deformities, and miscarriages and stillbirths were observed. It was confirmed that these health risks extend to mammals, since the same diseases were found in animal tests on rats, mice, and monkeys.

I would like to provide a general point with regard to animal testing: in the past, animal testing could certainly be justified. However, the current state of science, with its many diagnostic and therapeutic options, is such that we most certainly must give up animal testing (see p. 65).

My own studies with 57 children, discussed in chapter 7.2 and 8.3, seem to confirm these findings. Hence, we must assume these substances to be toxic to humans. In each of the examples so far, the people concerned were exposed to two or more metals.

Like mercury, tin is a precious metal and the two are very close in the electromotive series. Hardly any scientific studies exist on health damages in humans caused by tin. Once more, M. Daunderer (16., 17.) was one of the first to point out these risks. Hardly anyone knows that tin, especially in the form of organotin compounds, is about 1,000 times more toxic than inorganic metals.

Again like mercury, tin is removed from amalgam through corrosive and electrolytic processes. It is transported from the mouth, via the blood and lymphatic fluid into the bowel, where specific bacterial and fungal genera convert inorganic tin into organic tin compounds via a process called biomethylation. The process of transmethylation then creates the various organic compounds, such as di- and trimethyl tin.

The body maintains an equilibrium between noxious and endogenous substances (antigens and antibodies in particular). If a continuous feed of noxa, coming from, e.g., amalgam, disturbs this equilibrium, chronic diseases may develop. Daily, every single minute and second, toxic metal ions dissolve from amalgam alloys and are transported through the body via the blood and the lymphatic system. This is referred to as chronic poisoning by harmful substances, which in many circles people really do not like hearing.

Exposure to tin and its compounds can also occur through the respiratory system (in particular in the workplace) and through nutrition, with seafood consumption contributing 10 % to 30 %, depending on the particular contaminant and individual nutritional habits. In addition, tin compounds can be absorbed from clothing, as in the case of the soccer player from Borussia Dortmund whose jersey contained TBT. The latter case was extensively discussed in the media.

Organic metals can cross the blood-brain barrier more easily than inorganic ones—another fact that is unknown to many. I asked myself the following two questions:

1. What health damage can tin compounds cause by themselves?
2. Is it possible to conclude exposure to specific metals on the basis of the symptoms?

Answering the second question was effectively a sporting challenge to me, for which I coined the term metal-symptogram.

8.4.2. Method

In this study, I concentrated only on the health damage caused by inorganic and organic tin compounds. My control group was once again the 377 patients from my general medical practice. A total number of 84 participants from the group of 360 patients with environmental health damage showed exposure to only tin compounds.

Again urinalysis was performed by the Bremen medical laboratory (previously known as Schiwara laboratory, now operated by J.W. Wiitke & Co). (78., 125.)

8.4.3. Results

From the group of psychiatric diseases caused by tin compounds (both organic and inorganic) the following symptoms were mentioned most frequently: sleeping problems, fatigue, lack of motivation, inner turmoil, and general loss of performance. Most frequently mentioned neurologic diseases were headaches, vertigo, concentration problems, and nerve irritations.

From the organic tin compounds, levels of methyltin compounds were elevated by a factor of up to 15, compared to normal values and monobutyltin compounds even by a factor of up to 1,000. Levels of dibutyltin compounds were elevated by a factor of up to 3,000 and more. These extremely high levels of exposure were found particularly frequently in children with amalgam fillings.

When tabulating my data, I used the term tin-symptogram, which should express the fact that symptoms and clinical examinations by themselves can be enough to suspect exposure to tin.

Organic tin compounds by themselves can lead to the following health problems:

Table 2: Tin-symptogram: symptoms/diseases present after exposure to tin and organic tin compounds, as measured in empty-stomach urinalysis (total number of patients n = 84, number of patients in the control group n = 377. Important: for the control group I have not listed individual symptoms but only the number of diseases that could be diagnosed with certainty. For this reason, the fields listing symptoms remain empty.).

Symptoms/Diseases	Group of patients with exposure to tin and organic tin compounds		Control group	
	Number of patients (total: 84)	number in %	Number of patients (total: 377)	number in %
Fatigue	52	62		
Loss of motivation	40	48		
Pressure to perform	24	29		
Sleep disorders	29	34		
Total number of psychiatric disorders	145	173	30	8
Headaches	43	51		
Vertigo	25	35		
Concentration problems	32	38		
Nerve irritations	18	21		
Total number of neurological disorders	118	145	35	9

8.4.4. Critical assessment

The obtained results are remarkable and play a central role all through the study. Within my general practice, each symptom, and hence each disease, appears more frequently in patients with exposure to metals than in the non-environmentally exposed control group, with factors between 2 and 29. Like so many others before, this study shows that there is a clear relation between the frequency and severity of diseases on the one side and heavy metal exposure measurements from urinalysis on the other: the higher the load factors, the more severe the symptoms, which comes as no surprise to anyone who intensively occupies themselves with the topic.

The impression becomes even more striking when considering patients with exposure only to tin. The amount of neurological diseases increases to 145 %, in the case of psychiatric condition even up to 173 %, which means that each patient suffers from at least one or two simultaneous diseases of the nervous system and the psyche. It appears that organic tin compounds are many times more toxic than inorganic mercury: they cause even more damage to the psyche and to the nervous system.

To this we can add the high occurrence of skin and respiratory diseases, each appearing in 80 % of cases, which are a further indicator of tin's high toxicity. In fact, amounts of organic dimethyl- or trimethyltin compounds as low as a few nanograms are sufficient to cause severe chronic symptoms of poisoning and correspond to those caused by a simultaneous exposure to mercury, copper and tin.

The more than 3,000-fold elevated organotin exposure values that we found must be interpreted as an alarm signal!

> *Because of the grave medical implications of these observations, I appeal to all medical faculties to study these correlations in more depth.*

Limiting studies into amalgam to only mercury, as was commonly done in the past, does not provide us with a complete picture and can even lead us to wrong conclusions, which most certainly provides a further explanation as to why so far no conclusive evidence could be found for the damaging effects of amalgam.

> *The problem is not amalgam per se, nor is it limited to mercury, but it is the combination of all individual constituent substances that causes the harm.*

When observing the various organ disorders, the fact sticks out that we encounter many rare diseases of so far unknown genesis. Some examples of psychiatric illnesses are hyperkinetic syndrome in children and treatment-resistant depressions, panic attacks, anxiety disorders, and psychoneuroses in adults. Heavy-metal exposure was found in all these patients.

We can answer the second question, whether these symptoms merit a possible tin exposure, with 'yes'. For patients, especially children, suffering from fatigue, headaches, and respiratory and dermatological diseases, we may suspect exposure to tin if these symptoms occur all at the same time.

The obtained results show that there must be a causal connection between disease and heavy metal exposure. This is corroborated by the fact that these diseases occur more frequently in exposed patients than in the control group. In addition, there occur more frequent diseases in all organs in this group of patients. We must classify organic tin compounds as particularly toxic, especially to children, with amounts as little as a few nanograms sufficient to cause severe health damage.

In the future, we must strive to cease the use of heavy metals in dental restoration materials. In any event, new dental material must be thoroughly tested before being approved for use.

8.5. Environmental diseases in 360 patients with heavy metal exposure

Metals—the hidden causes of chronic diseases and benign as well as malignant tumors? (135.)

The studies summarized in the present text (130., 131., 132., 133., 134.) ran from 1990 to 2011. Trying to describe all work done on this topic would spring the bounds of this book.

8.5.1. Introduction

Studies conducted by W. Blumer (12.) showed for the first time that metals can cause cancer and chronic diseases. We now know that many metals have toxic and carcinogenic properties (7., 15., 16., 17., 30., 31., 58.). However, so far no measures have been taken. The creeping danger to our health posed by contaminants, fungal toxins, and radioactive as well as natural radiation continues unabated.

Regrettably, even today conventional medicine seems to either not see or outright ignore these new insights. Although over the last 30 to 40 years there has been a dramatic increase in incidence of environmentally related diseases, almost no research whatsoever has been done to clarify these questions.

In 1990, with this knowledge in mind and with results from my own research in hand, I started five studies. My research was helped by the fact that I had chronicled every case known to me of patients suffering from environmental diseases.

8.5.2. Method

The analysis for presence of metals was done by the medical laboratory in Bremen (previously known as the Schiwara Laboratory). During urinalysis, in order to better interpret the effects of diuresis[24], routine creatinine[25] measurements were taken. Depending on the measured levels, values were listed in g/d or µg/d.

Immunological tests using LTT (lymphocyte transformation testing) were performed by the medical institute in Berlin. Genetic characterization of enzymes known as Glutathion-S-Transferases was done by the institute for pharmacogenetics and genetic disposition in Langenhagen. In essence, these enzymes are responsible for heavy metal detoxification in the body, in particular for carcinogenic metals, cytostatic agents, and oxides as well as epoxides found in, a.o., cigarette smoke, drinking water, and pesticides. Therefore, these factors can be considered relevant to our health.

The first study assigned symptoms displayed by 360 patients with suspected environmental health damage and with heavy metal exposure determined from urinalysis to specific organ disorders and compared the findings with similar diseases in 377 patients visiting a general medical practice. The most frequently targeted organs were the psyche, the nervous system, skin, respiratory organs and the skeleton as well as connective tissues. Essentially, however, all organs were affected.

8.5.3. Results

Of the 360 patients with heavy metal exposure, 68 % had psychiatric complaints, 67 % displayed neurological symptoms, and 61 % mentioned skeletal pain and pain in the connective tissues. This is in contrast to only 8 % of patients from the general practice mentioning psychiatric complaints, 9 % having neurological symptoms and 28 % with skeletal and connective

---

24 Diuresis = excretion of urine
25 The effects of diuresis were taken into account by their relation to the concentration of creatinine, which allows unambiguous interpretation of the results of the analysis. Tabulating the creatinine measurements is outside of the scope of this work and for that reason I have not listed them. However, there exists a correlation between heavy metal exposure and creatinine clearance values.

tissue problems. These differences are significant and extend to dermatological and respiratory system disorders.

For patients with the following diseases, my studies found heavy metal exposure

Psychiatric diseases: depression, psychoneuroses, anxiety neurosis, panic attacks, hyperkinetive syndrome, vegetative dystonia.

Neurologic and neurodegenerative diseases: multiple sclerosis, Alzheimer's disease, Huntington's disease, gait disturbances of unknown etiology, progressive muscle dystrophy, migraines, macular degeneration, trigeminal neuralgia, hypersensitivity to chemicals, chronic fatigue syndrome[26].

Diseases of the skeletal and connective tissues: Primary chronic polyarthritis (PCP), fibromyalgia, hip dysplasia, herniated discs, polyarthrosis, respiratory diseases, allergies, hay fever, allergic asthma, recurring infections, chronic sinusitis, gastrointestinal disorders, ulcerative colitis, gallstones, hepatitis of unknown etiology, chronic pancreatitis.

Dermatological conditions: necrotic erysipelas, neurodermitis, psoriatic arthritis, chronic eczema.

Diseases of the sensory organs: Sicca syndrome, olfactory disturbances, tinnitus.

Cardiovascular diseases: Raynaud's disease, arteriosclerosis, varicosis.

Hormonal disturbances: Basedow's disease, thyroid adenoma with elevated mercury levels measured in the tumorous tissue, endocrine orbitopathy combined with fibromyalgia.

Autoimmune diseases: lupus erythematosus cutaneus and visceralis, diseases of the urogenital tract, prostate adenoma (benign tumor of the prostate), glomerulonephritis, uterus myomatosus (benign tumor of the uterus).

The causal connection between heavy metal exposure and the numerous diseases is supported by patients' medical history, clinical diagnosis of diseases, laboratory results, and last but not least successful therapeutic treatment.

None of the studied cases could be helped by conventional medicine, since no search after the causes of the diseases was undertaken.

---

26 You can read a summary of environmental pollutants and neurodegenerative diseases in H. U. Hill (52.).

My findings overlap those of M. Daunderer (15., 16., 17.) who pointed out the health risks posed by harmful substances, metals in particular, in numerous books and publications.

8.6. Heavy metal and harmful substance exposure in six cases of rare diseases of unknown etiology

In my second series of tests, I targeted testing for heavy metal and harmful substance exposure in patients with the following disorders: progressive muscle dystrophy, psoriatic arthritis, tinnitus, abortion[27] (miscarriages and stillbirths), endometriosis, and neurodermitis. These disorders are relatively rare, but during the past 25 years their incidence in my environmental-medical practice has increased sharply.

8.6.1. Progressive muscle dystrophy

Today, the causes of progressive muscle dystrophy are unknown. We now know that it is a hereditary disorder, but the cause of the underlying genetic damage is unknown. Since there are numerous differing forms of muscle dystrophy, each with different underlying genetic damage, we may assume a common cause, which hypothesis is backed up by the observation that metals and other harmful substances are capable of causing genetic damage. In my studies, I examined three patients suffering from progressive muscle dystrophy.

Results
All patients showed exposure to two or more heavy metals. I found mercury, copper, and tin two times for each metal, and gallium, cadmium, and cobalt once each. In one patient I found exposure to heavy metals and two chemical substances (PCB and HCH = hexachlorocyclohexane) as well as psoriatic arthritis.

8.6.2. Psoriatic arthritis

In the case of psoriatic arthritis too we still do not know its causes. Hence, it was not possible to initiate a targeted treatment. We studied a total number of nine patients.

---

27 abortion = miscarriage; lat. abortus = lead away from the mouth, abortion of foetus that weight < 500 g.

Results

With one exception, all patients showed exposure to heavy metals, with four of them being afflicted with three or more heavy metals. The one exception showed sensitization to tin and four patients had simultaneous exposure to PCB. In one of the cases, besides the triple exposure to heavy metals and the elevated PCB levels, a sensitization to gold was found. We didn't find exposure to harmful chemicals in any of the cases.

8.6.3. Tinnitus

By now, tinnitus is a widespread disease, to the extent that is frequently leads to professional disability. Therefore, the matter is of economic importance. We studied a total number of 13 patients.

Results

Twelve out of the thirteen patients had been exposed to heavy metals, two or more of them in nine cases. Only one patient showed concomitant exposure to PCB. One other patient showed exposure to two heavy metals and sensitization to silver and mercury. In one patient, we found hypersensitivity to silver, however no exposure was found.

8.6.4. Miscarriages and stillbirths

Any miscarriage or stillbirth is a grievous tragedy for the parents and their family, from which many mothers and fathers hardly ever recover. The case of multiple stillbirths is especially terrible and becomes a massive tragedy if the parents have specifically desired children.

It is regrettable that we as physicians try to find the underlying cause of stillbirths in only a rare few cases—in Germany, postmortem examination of stillbirths is uncommon. Hence, these terrible cases are seen almost as an act of fate not only by the parents, but also by the attending physicians. Laboratory tests on the mother or on the deceased neonate are hardly ever performed: there is no lobby for the dead.

Each death caused by diseases such as swine flu or measles is one too many, but each of these cases is published in the literature. In contrast, none of the responsible parties seems interested in miscarriages and stillbirths.

Results

Here too, my results show that as a rule we can find the causes for these cases, and hence can eliminate those causes. My study examined five mothers. All of them displayed exposure to heavy metals. One patient also had elevated levels of PCB and HCB (hexachlorobenzene). Both of these substances are highly toxic and are known for causing hormonal disruptions. Although this does not prove that these exposures caused the stillbirths, we can take it as clear evidence that these types of tests must be conducted in order to clarify the cause of death—they also confirm the results on intrauterine foetal damage, described in chapter 7.2.

8.6.5. Endometriosis

Endometriosis is a disease that is encountered quite frequently in en environmental physician's practice. Metals are capable of causing hormonal disruptions and damaging cells, so there is no reason to assume they cannot cause endometriosis either. My results appear to back up this assumption.

Results

Heavy metal exposure was found in all four patients with elevated levels for two metals in three of them. Two patients displayed exposure to harmful chemicals, three different substances (PCB, HCH, and DDE) in one patient and only PCB in the other. DDE is an abbreviation of dichlorophenylethane, which can be transferred to the infant through breast feeding, but which can also be transported to the foetus intrauterine.

8.6.6. Neurodermitis

All seven examined patients displayed exposure to heavy metals. Two of them had elevated levels of two or more metals. None of the examined patients showed exposure to harmful chemicals.

Aggregating the results above, we find exposure to tin and organic tin compounds to be most frequent, with 74 % incidence, followed by copper with 31 %, and mercury with 29 %. Further elevated levels were found for palladium, cobalt, platinum, gallium, thallium, cadmium, and gold as well as several heavy metals that are used in dental alloys.

As expected, PCB exposure was the most frequent of the harmful chemicals, with 21 %.

8.6.7. Critical assessment

With 74 % incidence, it is the highly toxic tin compounds, rather than mercury, that are most frequently found, followed by copper (51 %). These findings confirm prior results and provide further evidence that metals, in particular tin compounds, are at least partially responsible for the development of the disorders discussed in the previous sections.

The fact that in the 39 participating patients 97 cases of exposure to metals was found strengthens this observation. Strictly statistically, this corresponds to two metals found in each patient, with each combination potentiating the working of the other. The causal connection between exposure to metals and diseases have been confirmed during conversations with chemists and metallurgists and the issue is a known problem in both of these fields.

The instances of miscarriage and stillbirth are at the same time a sign and a warning: not only are children born with toxic substances in their body but—even worse—they are frequently killed by harmful substances whilst still in the mother's womb. The Munich University pathological institute conducted laboratory tests, conducted after the big amalgam trial against Degussa AG. These tests showed exposure to heavy metals in foetal brains.

As mentioned before, Degussa was not convicted of any crime but was required to pay a donation of 1.2 million German Marks, which money was used to fund further scientific studies. Hence, the fact of intrauterine damage to the foetus has been known since at least the time of this process. Nevertheless, no reaction has been seen and the results from these studies have, to my knowledge, never been officially released.

In addition to the heavy metal exposures measured during urinalysis, my studies revealed elevated blood levels of 13 further harmful chemicals. Here too, we see an interaction between the individual substances, which leads to mutual reinforcement of their toxic properties.

We now take a look at how these substances are distributed. If we look through the individual diseases, we notice that of the 13 cases of tinnitus, only one shows exposure to harmful chemicals. Hence, it is not far-fetched to conclude that metals in particular are responsible for causing tinnitus—

something which I could confirm through acupuncture and Interference Diagnostics, incidentally. All seven cases of neurodermitis displayed exposure to metals and not to other harmful substances. PCB appears to only play a causal role in the development of psoriasis.

For the sake of completeness, we must mention that the tests for harmful chemicals were only carried through for patients in whom we suspected such an exposure on the basis of their personal history.

Overall, the number of patients per disease is too small to draw unambiguous conclusions. Nevertheless, these findings give sufficient reason to research these connections within the context of a larger-scale study.

8.7. Exposure and hypersensitivity to metals in 139 patients with a benign tumor

8.7.1. Method

This study examined 139 patients with a benign tumor for exposure to heavy metals. 83 of these patients were also tested for hypersensitivity to metals. All patients had comorbid diseases and the benign tumor was a secondary diagnosis.

8.7.2. Results

For the participants of this fourth study, suffering from a benign tumor, exposure to on average two metals was found during urinalysis. For 46 % of the 83 patients that were tested for hypersensitivity, the LTT was positive.

The publications by D. Haase (46.), P. Jennrich (59.), M. Müller et. al. (87.), and G. Westphal et. al. (121.) on genetic damage caused by metals specifically induced me to research the simultaneous influence of metals and of immunologic and genetic factors on the development of malignant tumors.

8.8. Influence of heavy metals and immunologic and genetic factors on the development of malignant tumors

8.8.1. Introduction

Before presenting this article (132., 133.) I would like to tell a story. In the middle of the past century, W. Blumer (12.), a family physician from Netstal

(Glarus, Switzerland) noticed that cancer cases from his practice seemed to focus on the area near the main road that went through the center of the municipality. In 1958, he started a scientific study of 231 patients living directly on the main road, which lasted for 18 years. He compared these patients with those living closer to the municipality's periphery. We would nowadays say that he compiled a cancer register, such as the one first started in Germany, in Berlin.

Result of Blumer's research: near the main road, mortality rates due to cancer were seven times higher than for patients living in low-traffic zones.

It was immediately clear to Blumer that this observation was an extremely important new insight into cancer's etiology. At the same time, he found a higher incidence of other symptoms and diseases, such as headaches, gastrointestinal problems, fatigue, nervous disorders, depression, and prescription drug abuse in his group of test persons.

At that time, the lead content of automotive exhaust gases was still very high. Hence, Blumer assumed a connection between his patients' symptoms and this high lead content. In 1961, he started heavy metal detoxification treatment in 59 patients from his control group and studied mortality rates due to cancer for the duration of 18 years. It turned out that in this group, mortality rates due to cancer were around 90 % lower than in the group that did not undergo detoxification treatment.

This dramatic result first showed that heavy metals can play a significant role in the etiology of cancer.

These results were an incentive towards the following drastic reduction in the concentration of lead in automotive exhaust gases, which was an already massive step towards cleaner air. However, nothing further happened: a few years after Blumer's studies, his results and propositions had been forgotten and were left without follow-up.

Clearly, nobody considered metals and harmful chemicals other than lead to be dangerous to human health. One example is provided by diesel fuel, the constituents of which can be carcinogenic—a fact which has been known for years and against which even the best catalytic converters offer no protection.

Hence, my studies attempt to tie into Blumer's results, as it were, and intend to use the example of malignant tumors to put the dangers posed by metals back into the public awareness. For this, I followed unbeaten paths.

The following fact is shared by all prior studies: each one studied only one aspect, i.e., either heavy metals, genetic factors, or immunologic factors, as possible causes of cancer. In contrast, my pilot study aimed at gaining insight into the influence of all three factors together on the development of various types of malignant tumors.

For this, my study focuses on the role of metals in the development of cancer for the following reasons:

1. Presence of metals is easily determined in urine and tissue samples.
2. My own studies, as well as results in the literature show that metals play a particular role in the formation and development of malignant tumors. According to a list published by the US Environmental Protection Agency (EPA) and the Agency for Toxic Substances, these metals are many times more dangerous than formaldehyde or other chemical substances (59.).
3. The above rationale was also presented by P. Jennrich (60.) in his speech to the European Commission for Social, Health, and Family Affairs, held on the 15.11.2010. Metals are prevalent in daily life and accumulate in the human body with various toxic and immunological effects. Usual diagnostic procedures are incapable of diagnosing chronic metal exposure.
4. If we read the scientific literature and the latest DPA press releases, we find that during the past five years soil and air have been increasingly polluted with toxic metals. This also concerns plants and animals that serve as our sources of nutrition. All the responsible parties within our healthcare system have treated the creeping danger posed to our health by toxic metals as an ugly stepchild and it was even, to some extent, completely hushed up.

Forty randomly selected cancer patients participated in our study, with an average age of 68 years.

Six surgical departments from the Lüdenscheid clinic contributed to our study, as well as three different laboratories. We conducted multi element analysis for 14 heavy metals in patients' urine, in healthy organs, as well as in tumorous tissue. For analysis, inductively coupled plasma mass spectrometry (ICP-MS) was used. All measurements are listed in µg/kg. Since no reference values exist for tissue samples, all measured load factors are compared to normal urinalysis values.

Patients were tested for heavy metal sensitization for the same 14 metals, using lymphocyte transformation test (LTT). Genetic tests determined levels of the three Glutathione-S-Transferases GSTM1, GSTT1, and GSTP1. The studied malignant tumors were: ovaries, uterus, breast, lungs, pancreas, and colon (including the rectum).

8.8.2. Results

Finding the cause of any malignant tumor critically requires a thorough anamnesis that takes into account the patient's family history, as well as their professional and social history. In addition, anamnesis must also analyse any dental restoration material worn by the patient.

All patients wore amalgam fillings either during the interview or in the past. Of them, 93 % wore two or more different alloys and in 50 % of cases the patient's medical history pointed us towards suspecting professional exposure to heavy metals as a possible cause of the found elevated levels.

There was no significant difference in measured load factors between the individual organs for both tumorous and tumor-free tissue. There also appears to be no concrete connection between heavy metal exposure measured in tissue samples and the pathologic-histological diagnosis made. Hence, microscopic diagnoses do not justify concluding metal exposure to be the cause of malignant tumors. In fact, they could not point towards any particular cause at all.

For each patient, empty-stomach urinalysis revealed exposure to at least two to three toxic heavy metals, most frequently the carcinogenic cadmium and nickel. As many as five to six toxic metals were found over all the tissue samples, always including carcinogenic aluminium and nickel, but levels of zinc were also found elevated in these samples. Markedly elevated levels of carcinogenic cadmium were found in tumor-free tissue in 67 % of cases and in tumorous tissue in 58 % of cases, followed by cobalt with 38 % and 5 % in tumor-free and tumorous tissue, respectively. In 10 % of cases, lead was found elevated in both types of tissue (133.). The measured load values were significantly higher in the tissue samples compared to urinalysis, by factors typically in the two-to-three-digit range, on occasion even by a factor of a thousand and more. These findings are even more significant, since the subjects of the study were a randomized selection of cancer patients.

In 40 % of participants we measured an additional sensitization to metals, with total immunodeficiency due to chemotherapy occurring in three patients.

Genetic damage in the form of deletion (damage to the genetic material) and/or genetic polymorphism (variations in genetic frequency, changes to the DNA sequence) was found in 70 % of patients, with 51 % measured in GTSM1 and 18 % in GSTT1—these values are within accepted norms. Analysis of the Glutathione-S-Transferase GSTP1 tests showed 11 % variation of *A/*C or *B/*C, which is an almost threefold increase over the 4 % norm value. These results illustrate the role that genetic disposition plays in the aforementioned tumors.

With very high likelihood metals can be considered one of the common causes of genetic damage and immunodeficiency, which is supported by the extreme loads measured for carcinogenic and potentially carcinogenic heavy metals and the additional potentialization of their toxic properties and results confirm the data from the relevant literature. This all leads into a vicious circle: exposure to metals – Glutathione-S-Transferases – genetic damage – disruptions of the body's own metal detoxification systems and weakening of the immune system – further exposure to heavy metals.

S. Mukherjee (85.), in his book "The King of Diseases" (85.) cites Bert Vogelstein: "We can summarize the revolution that occurred in cancer research in one sentence: essentially, cancer is a genetic disease." Against this, I want to state my contrasting opinion:

> *On the basis of the results found in many hundreds of published scientific studies, as well as my own studies, I conclude that the essential cause of cancer is not to be found in genetic damage but in exposure to harmful substances, in particular to heavy metals.*

However, other chemicals, such as dioxins, furans, PCB, etc., as well as fungal toxins and radiation with multiple wavelengths, can damage genes and hence contribute to development of tumors. It is mostly not only one substance that is responsible for the emergence of a specific disease, but the combined effect of two or more substances. Essentially, it is irrelevant whether or not any elevated exposure levels are measured in patient's urine. Here, we recollect the "boundlessness" of norm values (67.), which, however, have their place as a basis from which to start discussions.

Seen this way, the developmental history of cancer has, as we will see later, a significant impact on diagnostic and therapeutic procedures.

8.8.3. Critical assessment

The participants in the first three series of studies had an average age of 43 years. We must mention that for the control group no thorough material analysis of any dental restoration material was done and no urinalysis for metal exposure was conducted, although all members of the control group wore dental prostheses. However, we were forced to allow this, since it was found to be impossible to assemble a sufficiently large group of patients wearing neither amalgam nor any other dental restoration material, which is—in and of itself—a medically significant finding: there are no adults without any dental restoration materials. Hence, there is a creeping danger of metal poisoning for a whole generation of people.

Each patient had a life of suffering behind them of on average 10 years, including numerous visits to doctors, before coming to my environmental-medical practice. In contrast, I had known a large number of the participants taken from my general practice for many years. With this in mind, the significant differences in the frequency of organ diseases in patients from the environmental-medical practice, as compared to those from the general practice, provide an additional indicator of the important role that metals play in the development of chronic diseases.

This fact is also confirmed by the results from the four series of studies: urinalysis showed exposure to on average two toxic metals in all patients. This cannot be coincidence.

> *Finally, the results of the therapeutic measures are proof of the correctness of the diagnoses: the diseases that were cured were caused by noxa, in particular by metals. Medicine can be so easy once we remove the cause of or the main factor in the development of diseases.*

The cure rate of 80 % for adults and up to 90 % for children testifies to the successful results of detoxification. These high rates of therapeutic efficacy are similar to the results found by the plaintiffs during the amalgam trial and to those found by the company health insurances in Essen who, between 1995 and 2002, in cooperation with the centre for the documentation of naturopathic treatments, conducted and analyzed detoxification treatments. After six months, recovery rates as large as 65 % could be shown (102.)

Regrettably, about 60 % of my environmentally damaged patients could not financially afford necessary additional dental restoration work. Here lies the true evil of the matter, to which I will return in chapter 10.2.3.

A further indicator of heavy-metal-caused immunodeficiency is provided by the high amount of hypersensitivity to metal in participants with benign (46 %) and malicious (40 %) tumors. It remains an open question as to why immunodeficiency causes benign tumors in some people, but malignant tumors in others. The group of patients with benign tumors was on average 43 years of age; in contrast, participants in the study into malignant tumors were on average 68 years old. Thus, the heavy metals had an average of 25 years more to wreak havoc on the organs of these patients.

Studies on eight patients with benign tumors of the thyroid (thyroid adenoma), prostate (prostate adenoma), or the liver (chronic hepatitis of unknown etiology) showed seven cases of exposure only to mercury. The patients were on average 50 years old.

In one of the patients with a prostate adenoma no elevated metal levels (including mercury) were found. Nine months earlier, at age 60, this patient underwent dental restoration and detoxification treatment, the last of which is repeated every year. These results also are sufficient to merit further studies.

There are numerous examples of benign tumors developing into malignant tumors.

Polyps in the colon and in the stomach are notoriously known for tending to become malignant after many years. The same is the case for gastrointestinal ulcers and other chronic inflammations. However, we still lack pathophysiological understanding of these processes.

We asked ourselves the question whether tumors are curable. The answer is that professional detoxification can prevent cancer. Professionally conducted holistic medical treatment can also cure these cancers on the condition that healing and prevention are considered specifically desirable from a sociopolitical and economic point of view.

8.9. Twelve years later: recognition of brain damage from environmental toxins

8.9.1. Introduction: why am I presenting this case?

Mr. B., 55 years old: severe depressive disorder from underlying brain damage caused by toxic substances.

I have decided to describe Mr. B.'s case in more detail, since it illustrates the fact that neurotoxic damage can imitate psychiatric symptoms so that even specialist physicians are unable to distinguish them from those caused by endogenous or acquired psychiatric disorders. In our case, Mr. B. had been diagnosed with psychiatric disorders for six years. Only after conducting environmental-medical examinations, biometric tests and imaging procedures, conducted during a further six years, a court recognized his disorder as a toxic psychosyndrome.

When it comes to the question of full or partial professional disability, legal proceedings typically take a lot of time. Expert witnesses for the public unemployment and pension funds like to diagnose patients with neurotoxic damage such as CFT or MCS syndrome only with psychiatric diseases, rating them as disabled according to their capabilities. Each time, the same pattern is followed: the public insurance funds commission expert testimony from neurologists, psychiatrists, psychologists, or experts on psychosomatic diseases, i.e., college lecturers, mostly professors from university clinics or medical academies. On first and second occasion, their findings are always the same, or at least highly similar: psychiatric disorders with or without narcissistic personality disorder, at borderline level, low therapeutic suggestibility, endogenous or reactive depression, significant chronification with nervous, often depressive, neurotic developments, psychosomatic disorder, and no chronic toxic damage to the CNS of the peripheral nervous system.

These primary diagnoses were also made in our case. Only later, it became clear that they were wrong. Our case study shows that these days it is possible to recognize virtually every chronic environmental disorder, in particular neurotoxic health damage. Beside conventional examinations, there exist numerous scientifically accredited imaging tests, as well as biometric and psychometric tests, and environmental-medical examinations (25.). It is necessarily advisable to conduct follow-up tests, which can find not only any worsening of the patient's condition, but also any improvement in their medical condition.

The case of Mr. B. described above is a typical example of the course of these types of expert witness procedures which, in this case, took more than twelve years to finally recognize his disease as toxic brain damage. Either intentionally or out of ignorance, no additional examinations had been conducted at any time.

Whilst describing this case, we will at the same time introduce the aforementioned diagnostic methods. Using those methods will enable the parties concerned in establishing expert testimony to present more scientifically recognized conclusions in the future, which may help get their claims recognized at a more rapid pace.

8.9.2. Some remarks to the patient's prior history

The anamnesis of the patient's family history showed nothing conspicuous. In particular, no diseases of the nervous system were known. In 1989, when Mr. B. was 36 years old, a neurologist diagnosed memory lapses as well as learning and concentration disorders followed, one year later, by severe psychiatric disorders, for which Mr. B. twice attended a private clinic in Southern Germany, followed by treatment in an outpatient clinic. Over the following years, he spent several extended periods in this clinic as well as undergoing ambulant psychotherapy at regular intervals.

Furthermore, Mr. B. complained about changing anxiety episodes and periods of agitation, sometimes even full-blown panic attacks. Come afternoon he was thoroughly exhausted, said he had reached the end of his mental tether and was incapable of doing anything. He found it increasingly difficult to be amongst other people as he felt under continuous observation, and he increasingly retreated into himself. He did not present any physical complaints other than occasional mild vertigo and a feeling of pressure in his head. When he suffered a panic attack he had occasional palpitations and shortness of breath. Additionally, I noted his highly sensitive olfactory apparatus and his intolerance to alcohol.

As the medical expert witnesses diagnosed his disease as not professionally related, we can neglect to mention his professional history.

8.9.3. Findings

From 1990 to 1996, electroencephalograms (EEG), echoencephalography, and visually evoked potential tests (VEP) were conducted at various times: all of them failed to find any pathologies. During this period, the numerous attending neurologists and psychiatrists made the following diagnoses:

- sensitive personality with neurotic developments
- personality disorder with sensitive characteristics leading to depressive syndrome.
- borderline narcissistic personality disorder with depressive syndrome
- neurotic developmental disturbance with a tendency to paranoid personality
- neurotic developmental disturbance
- disruptions to working capability with narcissistic structural components and contribution from a persistent adolescent crisis, with depressive syndrome

Since he was no longer capable of working, he qualified for social benefits for indeterminate time, early in 1993. Initially, payment of these benefits had been approved until 1996, but this was changed into an indeterminate period on the basis of an expert report by a neurologist from the LVA. This report essentially confirmed the anamnesis as well as the clinical findings so far. The expert witness did not find any neurological anomalies.

Psychiatric findings

"Mr. B. is a somewhat clumsy, sluggish person with avoidant, psychasthenic, narcissistic personality disorder suffering from occasional paranoid delusions. Despite his many years of psychotherapeutic treatment, he has gained little insight in his own psychodynamic circumstances. There is no significant organic brain damage. Occasionally, the patient suffers from depressive mood swings. Mr. B.'s disorders are strictly psychiatric, to the extent that we can speak of a borderline disorder, from which a neurotic and depressive disorder, accompanied by anxiety, developed. In these circumstances it is not considered helpful to conduct any therapeutic measures, as despite intensive therapeutic treatment, no future improvement is to be expected."

In addition, the expert witnesses considered Mr. B. to be disabled for work, a diagnosis with which he was not content. Although in 1989, his first attending neurologist found memory impairment, together with learning and concentration disruptions, no attempt had been made to clarify the underlying causes of these symptoms via further biometric testing and imaging techniques, although this finding should have been taken as an indication that the patient's disease was not strictly psychiatric, but also had an organic component. In the end, Mr. B. wanted answers to the following questions:

1. Can a causal connection be established between cerebral insufficiency and underlying neurotoxic damage?
2. Does this case concern an environmental-medical disease?

As little as eleven days after a psychiatrist diagnosed this case, Mr. B. underwent a brain-SPECT test, which showed pronounced damage due to blood circulation disorders. On the basis of these findings, he underwent numerous psychometric tests, imaging procedures, and environmental-medical laboratory tests during the next five years. In summary, these tests found clear indications of memory impairment, together with learning and concentration disruptions, as a consequence of acquired cerebral insufficiency, the cause of which presents itself in the form of damage to his cerebral glucose metabolism, as measured in PET tests. That Mr. B. had been complaining since late 1989 and the fact that the symptoms had been confirmed by specialist physicians justifies our assumption that the brain damage already existed at the time.

It is interesting to point out that at the time of the psychometric examinations (1999 and 2000) no indicators of any depression were to be found. On the contrary, the patient appeared very content with his life, from which we conclude that at no time he had suffered from a "primary psychiatric or psychosomatic disease". It is my assumption that him avoiding harmful and damaging substances contributed to his recovery—I have often found avoiding noxious compounds to be a precondition for long-term recovery.

Environmental-medical laboratory tests found exposure to DDE, a product of the degradation of the highly toxic DDT. These tests were conducted in the USA, as the norm values here in Germany are lower, and as a rule reference values from other countries are not recognized over here. In the case of Mr. B., due to the convincing nature of the other findings, the concerned parties made an exception—this presents a possibility that we could try to exploit during any expert witness proceedings.

8.9.4. The patient's application for rehab-measures

With the brain damage shown in medical imaging tests, as well as the environmental-medical findings, Mr. B. submitted an application for rehab-measures to the LVA Hannover. One concrete target was to obtain an official statement as to what type of work could be expected from Mr. B. and which measures are indicated to preserve, improve, or recover his capacity for work.

In addition, we sought renewed clarification as to whether Mr. B. suffered from toxic brain damage that could explain his disability for work as an environmental disease. Both this application and a next one were rejected: the patient's existing disability would not be mitigated by medical rehab-measures—in April 1998, Mr. B. appealed the verdict.

Once more, with the same rationale, the LVA's authority in charge of appeals rejected the application. However, in contrast to the first two official notifications, the LVA's expert witnesses now presumed two underlying conditions, one psychiatric and one environmental-medical. Literally, the statement said: "In essence, the appellant displays the following:

- severe narcissistic personality disorder at borderline level with limited therapeutic suggestibility and considerable chronification.
- toxic brain damage (encephalopathy) due to exposure to environmental pollutants."

An inflammation of the liver with comorbid disruption of the body's detoxification mechanisms was presented as a further diagnosis. In the course of the patient's disease, senior consultants were commissioned with preparing an ambulant neurological and stationary environmental-medical/psychiatric expert report.

The first of those senior reports was prepared in November 1999, when Mr. B. already felt better. The report concluded that at that time no noteworthy psychopathological findings could be established. Nevertheless, the report considered Mr. B. not eligible for therapy—and hence not capable of working—due to his personality disorder and his fixation on environmental poisoning. The consultant recommended a renewed examination by an occupational or environmental health physician with more experience within the relevant field.

This report was prepared in March 2000 and again, the expert concluded that no psychopathological findings could be established. Taking all previous findings into account, the report concluded that Mr. B. did not suffer from an MCS-syndrome (multiple chemical sensitivity), but from chronic fatigue syndrome (CFS). Literally: "The conducted psychometric tests can explain the patient's memory disruption as cerebral insufficiency."

An MCS-syndrome was ruled out and on the basis of the presented complaints, he suspected CFS, although at the time of the examination he could not establish its cause. He recommended to attempt CFS-appropriate rehab-measures to verify a possible environmental-medical cause of the syndrome. He concluded his report with: "Renewed PET tests, conducted in November 2000, again could not find any well-defined defects." The patient's memory disruptions were, however, not mentioned.

8.9.5. Critical Assessment

Mr. B.'s story is a classic example of how, in the past but also still today, public insurances and their expert witnesses judge neurotoxic damage as well as all MCS-related diseases as strictly psychiatric diseases, despite the availability of psychometric tests and medical imaging procedures as well as environmental-medical laboratory tests. The case we described is not a standalone case—I encountered many of them in my capacity of expert witness and attending physician for the patients involved. Now, the question as to the causes of these misjudgements necessarily appears, and we recognize them as numerous:

1. There is a high amount of ignorance among plaintiffs, expert witnesses for the insurance companies, attorneys, as well as judges. Already in 1994, the WHO classified MCS amongst "injuries and immunological diseases" in the International Classification of Diseases (ICD-1). According to this classification, MCS has been recognized as an organic disease (under the number T78.4) since at least 1994. The ICD categorizes psychiatric disorders in their F category (F00-F99). Hence, according to T. Metz (80.), the matter of possible psychosomatic causes had been settled before the "psycho-discussion" had been started. He continues: "Had the recognized state of medical science been taken into account, the psycho-discussion should never have taken place because, scientifically

speaking, the matter was decidedly settled at the time. However, neither plaintiffs nor the expert witnesses make use of this. Clearly they mixed up the state of medical science and the state of scientific discourse: the first has a legal definition and designates the accepted position and not the current position. It is solely the accepted position that matters here. Because procedures to change the official position take their time, the ruling position is of course dated. However, many patients and their attending physicians as well as expert consultants regrettably imagine this to be much different. Due to the trust they place in science, they rely on the so-called current scientifically accepted position in many areas even though they in particular should realize the extent to which part of science is influenced by various external interests. Hence, they forgot about and neglected legally accepted scientific standards. This is the wrong approach and damaging to the concerned patients.

In the context of MCS this means that we do not deal with "science guided by external interests". In fact, we really don't deal with science any longer, but with numerous expert witnesses misrepresenting the accepted state of science. The public can no longer see through these processes. As to why, we arrive at the next point:

2. There are only few environmental physicians and even less medical specialists (neurologists, psychiatrists) who have personal experience with environmental-medical diagnostic methods (environmental- and biomonitoring) and with access to appropriate biometric and medical imaging procedures. Hence, these tests are rarely prescribed. Even specialists such as neurologists and psychiatrists frequently refuse to send patients to another specialist (radiologists, neurologists experienced with psychometry) because they are convinced that the patient does not have any neurotoxic damage but rather suffers from a purely psychiatric disease. They are incapable of grasping the concept that neurotoxic brain damage caused by environmental factors can simultaneously cause severe toxic encephalopathy and/or peripheral polyneuropathy as well as severe psychiatric disorders. Frequent misdiagnoses are the consequence of this, which in our case could have been prevented. The memory and concentration lapses, olfactory disturbances, and finally the positive syndrome rapid-testing provided unambiguous evidence of neurological disorders

with organic causes. Hence, further examinations to clarify the causes of the diseases were urgently indicated.

3. A further aggravating factor is that public health insurance companies usually do not reimburse the costs of environmental-medical examinations. Costs for biometric examinations and medical imaging are only carried if they are ordered by a medical specialist. The patient, however, rarely is capable of paying for these tests out of their own pocket. Hence, it is not possible for them to find the exact causal connection between toxic substances and chronic neurotoxic diseases. Exaggeratingly, one could say that the economic boom of large chemical and pharmaceutical concerns comes at the price of no longer manufacturing toxin-free substances. Companies do not take responsibility for the health damage caused by their products, in our case DDE.

Hence, this case, together with the case of Matthias L., once again confirms M. Hennek who said:

> "The law tolerates many harmful substances. Our bodies, however, don't"

To make things worse, it is left to the patient to prove this causal connection. The health insurance companies' refusal to pay for necessary clinical examinations and tests really is incomprehensible from a medical point of view. After all, as we pointed out, all these methods are explicitly recommended by the relevant textbooks and represent current scientific opinion. Even the neurophysiological procedures have been known to us since the early 90s and are described in detail in N. Bierbaumer and R.F. Schmidt's handbook "neuro- and sensory physiology" (8., 9.).

What neurophysiological events happened in Mr. B.'s brain?
The cytotoxins that Mr. B. was exposed to led to a disruption of protein synthesis. This then resulted in an interruption of his long term memory, which remains even after blood perfusion improved.

4. There is an additional fourth cause that might play a role: state insurances and the expert witnesses employed by them consciously exploit these dubious cases to designate these victims as psychiatric patients, out of fear for a deluge of claims for damages. This way, it is all more simple and carries lower costs. This was admitted to me in numerous conversations with responsible parties from state insurances and health

insurance companies—something which they would never admit to in public.
5. However, the main reason is the fact that in Germany as in most industrialized countries today, environmental medicine is still neither taught nor researched in universities and medical academies as a subject in its own right. As a consequence, there are not enough qualified environmental physicians. Hence, insurance companies specifically select university professors who are only qualified in conventional medicine as their expert witnesses.

In light of the above, Mr. B.'s medical history is a classic case of a wrongful verdict, with the following particularities:

On the long and winding road that led from application to recognition, it took as much as 12 years (1989–2001) until his disease was recognized as toxic damage and hence as environmentally-caused. To the witnesses from the LVA the decisive factors were not the conventional medical findings, but the results provided by the numerous medical imagings and biometric tests as well as the biomonitoring. In this context we should point out that the LVA eventually recognized Mr. B.'s diseases as environmentally-caused, even though his levels of exposure to DDE and PCB were below the German legal thresholds. They were, however, above reference values in the USA.

Today, we still retain Paracelsus' proposition about the relation between dose and effectiveness, which states that the genesis and severity of a disease depend on the dose of the triggering substance. This proposition is, however, has been out of date for a long time. For example, if several noxa work together, they mutually increase each other's pathogenic effect. From this it follows that it is not just dosis and reference values of a specific harmful substance that are relevant to the causal connection between disease and noxa. Rather, we must take the number of different substances, the duration of exposure, the individual's tolerance for each of the substances, as well as the patient's age and gender into account in our causal analysis. This too, I pointed out multiple times and was confirmed ages ago in many scientific publications.

Further lessons that we can learn from this case:

> *Changes in the cerebral bone marrow only show up in MRT after many years. Hence, it is possible to prove the temporal correlation between the emergence of cerebral defects and the development of severe psychiatric symptoms. Similarly, we can objectify future improvements of the psychiatric symptoms through regression of the brain damage.*

8.9.6. My reasons for publicizing this case

I presented this particular case, because it is representative for numerous similar proceedings which conclude with more or less consciously made wrongful verdicts. I am fully convinced that today these verdicts are no longer justifiable from a medical and legal point of view, and hence are morally and ethically wrong. They are a nightmare for the affected patients, who already have a hard enough time coping with their chronic toxic health damage, cause them many sleepless nights, and if anything make them even more sick. In such proceedings the patient is more or less helplessly delivered into the hands of the expert witnesses employed by the state insurances, the doctors of the MDK (the medical service of the united health insurance companies), and, in the case of legal proceedings, the judges. This applies particularly when patients have to make a statement in a trial. In most cases this completely overstrains them.

> *Due to the difficulties in demonstrating proof, I recommend to ensure the diagnosis "neurotoxic health damage" as early as the application for disability or rehabilitation therapy by following the examination protocols that I described. This increases the chance of winning such cases and preventing protracted legal proceedings.*

8.9.7. Summary of the results of the studies

The results from the studies presented in this chapter confirm our urgent suspicion that harmful substances, in particular metals, are responsible for the genesis of chronic diseases and are (one of the) causes of cancer. The reason for this is that these diseases are a complex whole that is determined by multiple factors. Metals play a starring role because of their ubiquity, because we are in daily contact with them, and due to their increasing number. Hence, it is those metals that frequently cause the cup to overflow. Since their presence is easily shown in laboratory tests and because they respond very well to therapy, it is sufficient to conduct professional detoxification therapy in order to improve the patient's health and, in many cases, heal them of their disease. This kind of diagnostics is urgently needed in light of the by now huge number of patients with environmental damage, the unaffordable costs of our healthcare system, and the severe suffering of the affected patients.

On the basis of my own extrapolations, together with the numerous data in the literature, I estimate the number of people with environmental diseases at about 4 million. To this, we can add an estimated number of unrecorded cases as high as another 4 million. Since there are officially no patients with environmental health damage, there are no statistically sound numbers. An estimate 300,000 to 500,000 people suffer from multiple chemical syndrome (MCS) alone. Based on findings in a pilot study conducted by a special clinic in Bredstedt (Fachklinik Nordfriesland), a session of the Bredstedt society for patients with environmental damage estimated a number of 200,000–300,000 most severely diseases MCS patients. Besides members of the society, this session was also attended by a representative of the health insurance companies and Dr. Schwarz, head physician of the Bredstedt hospital (Klinik Bredstedt). The estimate was consciously set relatively low, out of fear that if a higher number was indicated, politics would immediately sign off.

The daily news from 03.12.2012 broadcast by ZDF (German public broadcaster) reported that the number of people with acknowledged psychological and physical disabilities alone is estimated at 8.7 millions. If one extends this number to the whole populace, we see that in Germany one out of every ten citizens has a disability. According to statistics collected by the Federal Statistics Office, released on 14.09.2010, 7.1 million people are severely disabled. Of this number, 64 % (5.6 million) involve physical disabilities.

Causes for these disabilities are, amongst others:

- congenital disabilities
- diseases
- accidents and injury during wartime or during military or community service

From this follows that, according to conventional medicine, harmful substances do not play any role as causes of congenital disabilities and diseases that accompany such disabilities. From an environmental-medical point of view this is incorrect, as shown by the case of the three-year-old Timo. If this were recognized officially, the number of people with environmental damage would probably be twice as large as the current estimate of three to four million, or possibly even triple. These are alarming numbers.

In the foreword to Orphanet's handbook (95.) we can read that about four million people suffer from one of the 7,000 rare diseases that we don't know the genesis of. Most of these are genetic disorders. If one takes these numbers as a basis, we must estimate the number of people with environmental diseases even higher than my own estimate of four to eight million.

Hence, it is necessary to study how many of these "rare diseases" and mental as well as physical disabilities are caused by exposure to harmful substances.

9. Environmental-medical diagnostics

9.1. Introduction: patients and their own interest in environmental medicine

In the preface, I quoted K. Müller: "Environmental medicine does not play any role in the public health policies of our political parties—including Bündnis 90/Die Grünen (German environmentalist party)".

The large number of patients with environmental health damage can be regarded as a direct consequence of this inappropriate policy. Because the subject is taught at hardly any of the German academies, there are simply not enough environmental physicians. Hence, it is unsurprising that even university professors are largely uninformed about environmental-medical analysis. We can ask and expect even less from private physicians and physicians working in clinics.

Yet, it all started out so well. From 1996 to 2000, regional medical associations still offered environmental-medical advanced training courses for their members to qualify as "Physician for environmental medicine". These courses were then canceled once more, apparently due to lack of interest. However, lack of interest could certainly not be the reason. In Westfalen/Lippe alone, each course was attended by about 100 doctors. This high participation rate testifies to the large interest that exists from within the medical profession.

The layman is also interested in the field. Via internet, friends, naturopaths, etc., they inform themselves about the influence that the environment exercises on their health. They consult their family physician trusting that their doctor will know everything—this is however not the case.

To this, we can add the fact that the public health insurance companies do not reimburse costs for environmental-medical services, or only to a very limited amount. Hence, environmental medicine in Germany is officially nonexistent. There is hardly any research on substances that have a potential for causing health damage, despite the dramatic rise in number of diseases of unknown etiology during the past 30 to 40 years. Some examples are neurological and psychiatric diseases, cardiovascular diseases, and cancer, as arrestingly pointed out by J. Mutter (89.) in his recently published book "Healthy Instead of Chronically Ill".

These negative facts stand opposite to the results obtained from evaluating the questionnaires issued by the KV, from which we see that patients are perfectly capable of recognizing sources, and hence causes, of their (potentially environmentally-related) disease, when asked for it specifically. The patient then expects clarification from his doctor.

To better understand the topic, I will use the following section to discuss the absorption of harmful substances by the human body. This will be followed by my description of the environmental-medical diagnostic method. This in turn depends on whether we are dealing with acute or chronic poisoning.

9.2. Internal exposure to metals

If we want to achieve lasting recovery from any disease, we must target the source of the malady. This source must then be dealt with in the same way that we deal with weeds, when we remove the whole plant, including its root.

In the context of this work, I will focus specifically on the absorption of heavy metals, harmful chemicals, and fungi toxins, which together account for the majority of poisonings. I do not claim completeness, but rather would like to present my own experiences. The procedure that we must follow plays an important role in reaching the source. This section is slightly simplified, in order to make it accessible to the layman.

How do I act? What are the symptoms? Which diseases are we dealing with?

Heavy metals
Heavy metals are mainly absorbed by the human body through:

- dental restoration material
- nutrition (fish, such as tuna), drinking water, contaminated soil
- air pollution with fine dust
- radiation (radioactive and nonradioactive)
- breathing in interior spaces (metal industry, galvanics)
- the skin, through clothing and coins
- toys and cooking utensils made from plastic (with particular mention of organic tin compounds)

- implants
- prostheses (endoprostheses)

Due to their large apparent danger, I would like to describe dental restoration materials in some more detail.

From my own experience, which is supported by numerous clinical studies and by the relevant literature, it is metal-containing dental restoration material that most frequently causes health damage due to exposure, and it is the most dangerous source. They should be considered as the most important source of diseases. There is a lot of truth in the title "Healthy Teeth, Healthy Human", that J. Lechner gave his book (71.).

As I already described before, we must also look at the development of cancer in connection with dental restoration materials. Hence, material analysis should always be part of a clinical examination.

Metal alloys in our teeth are more harmful to our health than occupational exposure to harmful substances. Why is this?

Metal ions are released through abrasion, evaporation, corrosion, and electrolysis, a fountainhead that bubbles 24/7, so to speak—day and night, week after week, year after year, and even when we are on a holiday. Imagine that you were electrocuted day and night or that you were forced to live in a room with glaring lights shining without any break. Our body does not get any restorative period to recover from its fight against continuous inflammatory stimuli.

Exposure to harmful substances in the workplace, at home, or outdoors, in contrast, is always interrupted by pauses during which the body is not exposed. Nevertheless, we should not neglect these substances, because their number has markedly increased over the last 20 to 30 years, to the extent that it has become almost unmanageable. Only 3,000 out of many millions of chemicals (nobody knows their exact number) have been tested for toxicity. One might say that chemicals as pathogens are becoming big movers and shakers. At the same time, their interaction with metals leads to a mutual increase in their harmful action.

Substances are absorbed by our body from the blood, through the skin, from the gastrointestinal tract, or directly through mercury evaporation rising toward the head and into the brain. Most people don't notice anything. We can't smell nor otherwise feel them. Only in rare cases do we notice a metallic taste in the mouth.

Until a few years ago, absorption of harmful substances through our nutrition was regarded as insignificant. Nevertheless, the WHO (World Health Organization) declared it to be the major cause of exposures, despite the literature having singled out dental restoration material, in particular amalgam, as the most important cause years ago. In the meantime, the WHO updated its stance despite the fact that, paradoxical as it may sound, absorption of harmful substances from our food also has increased markedly over the past 10 to 20 years. Since the production of amalgam and the usage of wildly differing dental restoration materials in general have increased even more strongly, metal-containing dental restoration material remains the number one source of danger. Politicians and the responsible parties within our healthcare system now seem to wake up.

> *Fundamentally, all heavy metals and all dental restoration materials can become toxic, once the dose and duration of exposure has exceeded certain thresholds.*

Which alloys and which heavy metals are we dealing with?[28]
The most important alloys:

- Amalgam: contains copper, silver, tin, mercury, and traces of nickel and palladium, the last two of which are not required to be listed by the manufacturers. Concretely, 50 % mercury is blended with identical amounts of a mixture of copper, tin, silver, and mercury. In the past, it was possible to observe this blending procedure during a visit to the dentist.
- Gold alloys: contain gold and platinum, and some trace amounts of hardening metals, such as indium.
- "Budget gold" alloys: contain gold, platinum, and a sometimes high amount of precious metals such as palladium. This alloy is cheaper than gold, which gives it its name "budget gold". Palladium is the most abundant naturally occurring metal, which makes it so cheap.
- Precious metal alloys: usually also contain up to 70 % (and more) palladium, together with copper, tin, silver, and/or other precious metal. They always contain nickel residues.

---

28 The following also does not aim for completeness. Its goal is to be an introduction comprehensible to the layman.

- Metallic prostheses: contain the carcinogenic metals cobalt and chromium, and molybdenum.
- Implants: most frequently are made from titanium. They are surgically fitted into the jaw and hence are in even closer contact with tissues and blood circulation. More recently, after it was discovered that titanium can cause allergic reactions, implants are also being made from zirconium. Like titanium, zirconium is a heavy metal. It is supposedly non-toxic, although this fact is contradicted by the positive cases of testing for hypersensitivity to titanium and zirconium (positive LTT). Thanks to these tests, I could warn the patient against an implant before any harm was done. After detoxification, they tolerated the titanium well and their immune system was back in good shape. In all cases, patients had amalgam alloys replaced with gold or budget gold alloys, without detoxification treatment. Although there were short-term improvements to their health, their situation eventually deteriorated again (see section 7.2.1). These cases are not exactly rare. After this happened, it was decided to place a titanium implant. As a consequence, patients' symptoms became worse. Had they been professionally detoxified, patients most certainly would have tolerated the titanium implant well.
- Endoprostheses: contain a top layer made from titanium. The prosthesis as a whole contains nickel, chromium, and iron, i.e., toxic and partially even carcinogenic metals. If abrasions happen, that is if the top layer starts to wear, these precious metal ions also enter the body through the bloodstream or the lymphatic system. I have diagnosed several acute and chronic titanium allergies that emerged after placement of an endoprosthesis—something which many surgeons and orthopedists do not want to be true. In the end, the endoprosthesis had to be exchanged after all.

After British publications in the scientific literature, from summer 2012, implants and endoprostheses have now also become publicly known as sources of metal poisonings.

What applies to heavy metals must be considered for non-metallic dental restoration materials as well. Concerning these we have even less experience.

9.2.1. Nonmetallic materials (plastics, ceramics, endoprostheses)

This group contains the synthetic materials (plastic teeth, prostheses made out of plastic), cement (glass ionomer cement), and ceramics. Ever since the dangers posed by metal alloys have become known, these substances have become increasingly more relevant. However, because these materials also are foreign substances to the human body, we can also expect them to cause health damage. Every single synthetic material should also be tested for tolerance with LTT before being introduced in the human body. This should be preceded by detoxification treatment.

Regrettably, today there are too few clinical studies on the tolerability of dental restoration material. Due to their variety—an estimated number of 3,000 different materials in Germany alone—such studies will also be hard to carry out in the future, in particular because these substances are often mixed together or used in conjunction with heavy metal alloys (see chapter 9.8.3.).

9.3. External exposure to metals

9.3.1. Nutrition, drinking water, soil, and air

Harmful substances are ubiquitous wherever we go on this earth. They are even transported to the arctics via water, rain, and clouds. Hence, it should come as no surprise that here in our industrialized countries we absorb those substances through our nutrition and our drinking water.

Strawberry fields are located directly next to a heavily used main road. Nevertheless, numerous citizens continue to harvest them when it's strawberry season. They are, after all, cheaper than in the supermarket, and at least they are plucked by hand. Maybe the ones bought from the store contain even more harmful substances.

Vegetables, fruit, berries—every single one of those may contain harmful substances not only due to air pollution, but also through contaminated soil. The same is true for our drinking water, which is tested for germs on a regular basis but not for toxic substances.

9.3.2. Clothing, toys, coins

Absorption of harmful chemicals via the skin and the respiratory system most frequently occurs in the workplace. Poorly ventilated rooms and absence of appropriate air vents stimulate exposure in interior spaces and hence stimulate absorption through breathing or via the skin. There are, however, other sources of exposure to harmful chemicals, which are largely unknown to most of us.

From 1960 to 1984, Polychlorinated biphenyls (PCBs) were used in sealing joints and neon tubes installed in newly erected schools, kindergartens, and other public buildings. Jürgen Jäger, representative for environmental and pollutant affairs from the Hessen Science and Education union (Gewerkschaft Erziehung und Wissenschaft GEV) estimates that PCB-containing materials were used in 15,000 out of 45,000 schools nationwide. One can easily imagine that a very large number of students, teachers, and other school staff have been and still are exposed to PCB ever since. This is, however, denied by officials, who fear a deluge of claims for damages.

We should also not underestimate the absorption of noxa through smog. Officials still pay too little attention to the danger to public health posed by fine dust. Aside from instating danger areas for trucks and cars and the installation of catalytic converters, which usually don't do as much for reducing harmful chemicals as they were promised to do, not many concrete measures have been taken. Example: we know that diesel fuel in particular contains carcinogenic substances, but yet it has not been banned.

Ever since Chernobyl and Fukushima, it has become clear that the public health danger posed by radioactive radiation has strongly increased during the past 20 to 30 years. Finally, the politically and economically responsible parties seem to have understood this fact, as testified by Germany's about-face with regard to energy supply. This recognition is rather late, but it might not yet be too late. The fact that the norm values (reference values) for this type of radiation have been raised shows, however, that they are not yet prepared to accept the ultimate consequence—instead of lowering the norm values, they have been raised. Thus, the individual will find it even more difficult to present a proof in case of health damage.

9.4. Interior and exterior exposure to other industrial products and fungi toxins

The rise in harmful substances in clothing and toys that are manufactured in low-wage countries is alarming. Usually, these products contain highly toxic organic tin compounds (butyl- and methyltin compounds). This development particularly affects toddlers and children. It is also incomprehensible that today we still make two-euro-cent coins that contain carcinogenic nickel.

Absorption of harmful chemicals
The sources from which we absorb harmful chemicals, also called industrial products, are largely identical to those for heavy metals. Chemicals are, however, usually more volatile and are more easily metabolized. They do not "besiege" our body day and night, like metals do. Nevertheless, they can be highly toxic. Examples of such chemicals are: polychlorinated biphenyls (PCBs), furans, dioxins, dichlorodiphenyldichloroethylene (DDE), hexachlorobenzene (HCB), lindane, numerous wood preservatives, formaldehyde, etc. We can extend this series indefinitely.

Fungal diseases
Wherever there are humidity and heat accumulation, fungi proliferate. At home, this applies in particular to insufficiently ventilated rooms. At work it occurs wherever we are dealing with insufficiently removed contaminated air. Fungi are absorbed preferentially via the respiratory system, the gastrointestinal tract, and the skin. In this case, the fungi toxins are the ones that pose the highest danger. In spring 2013, aflatoxine, a fungal toxin, was found in sources of nutrition (corn and milk). Radio, television, and the press spoke of a fungal-toxin-scandal. Aflatoxins are highly poisonous and carcinogenic. Once again, the official word was that there was no immediate danger to public health. The chronic effects caused by these toxins also went unmentioned—after all, they may have been circulating for months or even years. I am sure nobody actually tried to check this. We may make a mental note of the fact that fungi toxins are, in essence, capable of causing symptoms similar to those caused by all chemical substances.

9.5. The importance of a patient's prior history and of clinical examinations

The first and most important precondition for finding the cause of an environmental disease is recording an extensive patient history (anamnesis). This process can be made a lot easier, if the patient would write down his medical history before visiting the doctor, in particular when they have been sent to an unfamiliar medical specialist. The anamnesis should include family, personal, professional, and social history, but it should also contain information on domestic situation, location of the patient's house (is it located near a high-traffic road or a side road? Is it in the outskirts of a town or in the country?) Additionally, recording a description of the patient's workplace is important. Do they work in shifts? Which hobbies does the patient pursue? Finally, the patient's nutritional situation, eating habits, and stress factors in the workplace as well as in their marriage or other significant relation must also be recorded.

> *One should look into oneself, so to speak, and ask: can it be that my lifestyle is an additional, or even the main cause of my illness?*

One should not always immediately lay blame at the feet of others, but first look at oneself. When following this protocol, the success rate for finding the correct diagnosis is, as mentioned before, around 90 %. The list of symptoms and diseases should help the patient when preparing his prior history. With this self-anamnesis in hand, the patient can go to their family physician (see also the protocol described in chapter 12).

To the anamnesis should be added a thorough clinical examination. Today, however, there is hardly any time for both of the aforementioned medical necessities and hence they usually are skipped.

When making a diagnosis, we must discern between acute and chronic poisoning.

The anamnesis should include the following subjects

We should ask in detail after the following possible sources of exposure and vectors of absorption: nutrition, skin and clothing, interior exposure (air pollution, increased outgassing). We must also conduct material analysis of any dental restoration materials, which should include preparing a dental organogram.

I will come back to this in the next chapter and explain what we mean with a dental organogram. During anamnesis, we should also ask after possible dental implants as well as the presence of endoprostheses. Hence, the patient's prior history and conventional medical examinations should always be the starting point of the investigation.

This is for two reasons:

1. Conventional medical examinations are still the basis of any type of medical diagnostics. It would be a medical blunder to neglect conducting these tests.

2. If the case involves expert witness procedures, the patient's only chance to get their environmental disease recognized is when everything has been clarified from a conventional medical point of view. Only then one could, so to speak, prove with 100 % certainty that the disease was caused by harmful substances from the environment.

9.6. The difference between acute and chronic poisoning

Acute poisoning
Acute poisoning is easy to recognize, due to the spatial and temporal correlation between cause and disease. This connection is often easily found by asking the patient, who often will already point it out by themselves.

Chronic poisoning
Chronic poisonings have the following characteristic properties:

- as a rule, victims are unfamiliar with various types of poison and their action. Hence, they cannot avoid being poisoned, in a timely manner
- often, the concentrations involved are so small that no acute health damage is done
- those with prior health damage, as well as particularly sensitive people (infants, children, pregnant women, elderly and/or diseased patients) are the first to be damaged by poisons
- the effects of the poisoning only appear years after the fact
- often, more than one poison is involved. Each of them may have unknown interactions with any of the others, which makes toxicologic evaluation of the case impossible.

In Germany, as in many other industrialized countries, patient and attending physician must present 100 % proof of a causal connection between poisonous substances and the patient's disease.

9.7. Symptoms and chronic diseases due to exposure to harmful substances

Frequently, the same term, e.g., "headaches", is used for both symptoms and diseases. Neither of the two lists that follow claim completeness. Nevertheless, I think that every patient with an environmental disease will recognize their own symptoms. As I already sketched in the introduction, I am consciously describing symptoms and diseases separately from each other, so that the reader—the patient—can find out which symptoms belong to what disease. This way, after determining his own medical history, the patient will be better able to classify their complaints.

Symptoms

General symptoms
Headaches, malaise, concentration problems, increased fatigue, sleeping problems, vertigo

Psychiatric symptoms
patient is easily fatigued, lassitude, increased irritability, emotional lability, aggression, depressive mood disorders, problems with concentration and short-term memory, inner turmoil, reduced libido

Neurological symptoms
headaches, concentration problems, memory lapses, mental impediments, numbness (paraesthesia), itching, ataxia (gait disorders)

Internal/immunological symptoms
infections of the urinary tract, kidney function disorders, nose and throat infections, reddening of the mucus membranes in the throat and the tonsils, bronchitis, bronchial asthma, pseudocroup, fever attacks of unknown etiology, diarrhea, intestinal bleeding, recurrent bowel mycosis, damage to the spleen and lymphatic nodes, liver diseases

Hormonal disturbances
disturbances in the menstrual cycle, fertility problems, thyroid dysfunction (infections); overall, hormonal disturbances occur more frequently in females than in males.

Sensory organs
impaired vision and hearing, teary eyes, frequent conjunctivitis, disturbances of the taste buds and the olfactory nerves

Skin
Hair loss (alopecia), acne (chloracne)

From this multitude of symptoms we can gather that once the body absorbs harmful substances, they can be transported through the blood and lymphatic symptom to infiltrate every single one of our organs. Hence, a thorough anamnesis usually points directly towards the diseased organ. We now ask ourselves which organ diseases there are. Which diseases hide behind those symptoms?

Chronic diseases

Psychiatric diseases
Depression, panic attacks, hyperkinetic disorder, anxiety attacks, anxiety neurosis

Neurological diseases
Trigeminal neuralgia, toxic polyneuropathy, migraines, multiple sclerosis, toxic brain damage, Huntington's disease, MCS and CFS, progressive muscle dystrophia, Alzheimer's disease

Diseases of the sensory organs
Sicca syndrome, olfactory disturbances, tinnitus, macular degeneration

Diseases of the skeletal system and the connective tissues
Chronic rheumatoid arthritis, fibromyalgia, hip dysplasia, osteoporosis, arthritis, skeletal and connective tissue disorders related to psoriasis (psoriatic arthritis)

Gastrointestinal diseases
ulcerative and mucous colitis, Crohn's disease, hepatitis of unknown genesis, chronic pancreatitis

Respiratory disorders
allergic asthma, recurrent infections, chronic sinusitis

Endocrine disorders
primary sterility, habitual miscarriages, increased male body hair growth (hirsutism), masculinization (androgenization), struma, Graves' disease, endometriosis, ovary infections

Diseases of the blood and the lymphatic system
increased white and red blood cell count

Cardiovascular diseases
Blood perfusion disorders in the fingers, cardiovascular disturbances with vertigo attacks, tachycardia, blood pressure crisis, varicose veins

Autoimmune disorders
lupus erythematosus, scleroderma

Skin diseases
eczema, flaky skin, neurodermitis, skin allergies, pus-filled blisters on the skin (acne pustulosa), chronic psoriasis with involvement of the skeletal and connective tissue as well as the joints

Respiratory diseases
asthma, chronic sneezing, chronic sinusitis

Urology and gynaecology
irregular menstrual period, vaginal mycosis, inflammation of the bladder

Internal/Immunological symptoms
Urinary tract infections, kidney function disorders, nose and throat infections, reddening of the mucus membranes in the throat and the tonsils, bronchitis, bronchial asthma, pseudocroup, recurrent bowel mycosis, damage to the spleen and lymphatic nodes, liver diseases

Skin diseases
alopecia (hair loss), acne (chloracne), tendency to develop mycosis (27., 28.)

Benign and malignant tumors
From this multitude of symptoms we can gather that once the body absorbs harmful substances, they can be transported through the blood and lymphatic symptom to infiltrate every single one of our organs.

With this knowledge about sources and absorption of harmful substances, as well as the symptoms and diseases caused by them, I will now turn to the topic of case history.

Fungal diseases
Symptoms and diseases caused by fungi are little different from the ones described above. Fungi should always be considered if no other causes for a patient's disease can be found, or if an examination of their house or workplace points toward them as possible cause.

Method
Obviously, all standard conventional-medical tests, such as laboratory values, X-ray and CT (computer tomography) scans, MRT (magnetic resonance tomography), ultrasound sonography, ECG, EEG, etc., must

be part of the arsenal of environmental-medical diagnostics. Specific environmental-medical examination of the patient (environmental-medical tests and biomonitoring) should only be performed after these primary methods found no results, and should be conducted by a doctor with the relevant experience

9.7.1. Examination for external exposure (environmental monitoring)

These examinations involve measuring concentrations in the environment, in order to prove or exclude exposure to harmful substances. The goal is to identify the source of the exposure. These tests are conducted by environmental engineers or by institutes for interior diagnostics. If such an exposure is expected, one should contact these specialized professionals (see the Appendix).

This type of diagnostics tests for environmental noxa in interior air, house dust, suspicious domestic objects, building materials, drinking water, and foodstuffs. House dust is particularly suited to this type of testing, because it is capable of absorbing many harmful substances from the interior air, due to its large surface area, long retention period, as well as its distribution over and circulation through all of the interior space. Hence, house dust examination functions as a screening method.[29]

Vice versa, house dust and interior air can be tested for suspicious foreign substances right at the beginning of the investigation. In this case, the next step will be to test the people concerned for possible exposure to the found substances.

Domestic objects: furniture (due to wood preservatives), leather (may contain PCP), textiles (may contain formaldehyde), and carpet (once again, PCP) are the most frequently tested domestic objects.

---

[29] In the screening method, house dust collected from interior spaces is tested for presence of suspicious substances. If the found substances are identical to those found in the body, proving a causal connection between disease and domestic situation becomes a lot easier. The same of course particularly applies to the workplace.

Building materials: of particular relevance as sources of exposure are extensively worked wood- or woodchip-made plates (wood preservatives and formaldehyde, respectively).

Water: water contamination with noxa from the environment is found for example in cases of faulty sewage drainage, in particular industrial sewage, which is then deposited into rivers and brooks. In cases of high-tide catastrophes, large bodies of water infiltrate factories and are contaminated by harmful substances stored and used in those factories. Pesticides sprayed on farmland are transported into the groundwater and from there into brooks and rivers.

Foodstuffs: exposure to harmful substances coming from foodstuffs is becoming increasingly common. Sources of exposure are meat, eggs, plants, and by now even milk. Contamination is mostly caused by non-compliance with hygiene regulations and the failure of regulatory bodies. Frequently, such cases are made public, which makes it easy to clarify the causes of the contamination. Regrettably, over the last couple of years cases that are only discovered at a very late time are accumulating. In these cases, health damage may have occurred that is hard to causally clarify. In these cases, the plaintiff almost always pulls the shortest straw.

Reference values are important to assess situations like that. Here, external exposure serves as a marker. In order to evaluate the measurements, we make use of the so-called threshold value, called ADI, which stands for "acceptable daily intake".

9.7.2. Examination for internal exposure (biomonitoring)

This type of testing can be done by any doctor, in collaboration with an environmental-medical laboratory. Tested are the blood, urine, hairs, teeth, tissue, and eluate collected from apheresis (see also chapter 8.2.2.).

Depending on the patient's medical history and the clinical findings, the following environmental-medical laboratory tests are available:

1. urinalysis: testing the empty-stomach urine for presence of heavy metals, and the DMPS provocation test
2. chewing gum test (saliva testing)
3. hair analysis
4. tissue testing

5. dental testing
6. breast milk testing
7. analysis of eluate collected from apheresis or dialysis

Urinalysis to test for presence of heavy metals
In this test, (empty-stomach) urine is tested for presence of heavy metals after administration of DMPS (so-called provocation testing). From the patient's prior history and from material analysis of any dental restoration material, we can determine for which metals we have to test.

DMPS provocation testing, also mobilization testing
DMPS is the abbreviation of 2,3-Dimercapto-1-propanesulfonic acid as a sodium salt. It binds to heavy metals with decreasing affinity (binding capacity) and forms water-soluble complexes. This test is suited for mobilizing heavy metals in so called organ stores. Hence, it can be used to determine the deposit of heavy metals in body stores such as the liver, kidneys, central nervous system, lymphatic system, thymus, etc. For this reason, the DMPS test is also called a "toxicological magnifying glass". (16.)

Remark
Metals can, in principle, infiltrate every single organ, and accumulate. There are not yet any reference values for organs and tissue. If, however, the values increase thresholds determined for urinalysis or blood levels and if the patient's prior history and lab results point towards exposure to heavy metals, we may suspect this type of exposure on the basis of these values found in DMPS testing.

Mercury absorption rates from amalgam fillings may be several times higher than those from food (104.). Hence, DMPS testing is particularly suited for demonstrating chronic exposure to metals, which often can only be recognized by increased heavy metal secretion in the urine after administering DMPS and can not be determined from concentrations in the blood or in non-provoked urinalysis.

DMPS Testing procedure

1. Urine sample 1.: 20 ml empty-stomach urine, unprovoked, before administering DMPS (Dimaval), to be tested for presence of heavy metals
2. 3 mg DMPS (Dimaval) per kg of body weight, orally, in the form of a capsule, on an empty stomach

3. Drink 150 ml water
4. Urine sample 2.: 20 ml urine 45 Minutes after administering DMPS intravenously or 2 hours after oral administration, to be tested for presence of heavy metals.

Important: when administering DMPS intravenously, the injection should be given very slowly. I recommend administration via an intravenous drip (diluted with physiological salt solution, or with the patient's own blood, drawn into the syringe prior to the injection).

Side-effects: 1 % of patients may develop transient skin reactions after intravenous administration of Dimaval. Persistent vegetative and labile patients may pass out. For this reason always administer the injection slowly, or provide DMPS in a drip.

Contraindications: reduced kidney function and hypersensitivity to DMPS

I recommend that the patient remains in the practice for observation for up to an hour after giving the injection.

Sample tubes should be obtained from a specialized laboratory. Of course it is also possible to determine levels of heavy metals in the blood, but this is a much more elaborate procedure, and unsuited to in particular children.

For "only amalgam and no other exposures" I recommend urinalysis to test for presence of mercury, tin, copper, silver, and zinc. Additionally, the urine should be tested for presence of organic mercury compounds (ethyl and methyl mercury), and organic tin compounds, also known as organotin compounds.

This group of compounds contains:

- monobutyltin
- dibutyltin
- tributyltin (TBT)
- monomethyltin
- dimethyltin
- trimethyltin

Patients who wear gold alloys and/or other precious metal alloys should be tested with multielement analysis (MEA) and be tested for presence of organic metal compounds. With this method, we cover all heavy metals that are in use in dental alloys, which are in essence the following: mercury, copper, tin, silver, lead, nickel, aluminium, gold, palladium, indium, thallium, iridium, arsenic, cobalt, molybdenum, chromium, gallium, ethyl and methyl mercury (organic mercury), organic tin compounds

Chewing gum test (saliva testing)

The term amalgam means an alloy that is composed of the metals mentioned above. Because it contains mercury, which evaporates at room temperature, amalgam is, strictly chemically speaking, a "viscous liquid".

Aside from evaporation, metals can also be released from amalgam fillings due to corrosive processes. These processes are influenced by the patient's oral hygiene, the quality of the amalgam fillings, pH of the drinks the patient consumes, and mechanical stresses from chewing and brushing teeth. Hence, elevated levels of amalgam's main constituents (mercury, tin, copper, silver, gold, palladium, nickel, cobalt, chromium, etc.) can be measured in human saliva.

> *That which is valid for amalgam applies, in principle, to all metal alloys. There exist no dental fillings that do not corrode. What makes amalgam special is that we can add to the corrosive processes the fact that mercury evaporates from it at room temperature.*

By now it has been established that there is a correlation between mercury concentration in a patient's saliva, after chewing gum and mercury secretion in the urine after administering DMPS on the one hand, and vaporous mercury evaporating from amalgam fillings and excreted via exhalation on the other hand.

Since mercury secretion in the urine after administering DMPS and the concentration of toxic vaporous mercury in exhaled air are a measure for exposure to mercury, the chewing gum test is recommended as a screening test. Regrettably, there are no unambiguous threshold values. Hence, interpreting measurement results is quite difficult.

Method:

- No food for at least two hours prior to the test. Drinking is allowed.
- Collect 5 ml saliva in test tube I.
- Chew gum for the duration of 5 to 10 minutes, making sure that the amalgam fillings are participating intensively in the chewing action.
- During chewing, collect saliva in test tube II.
- Submit saliva samples I and II to laboratory testing.

Hair analysis
Hair analysis is very important in order to diagnose a patient with "drug abuse". However, it is increasingly used to determine heavy metal poisonings that happened further back in the patient's history. Hair analysis too is suitable as a screening test.

Examination of tissue samples
As mentioned before, there are no reference values for harmful substances in human tissue. This is highly surprising, since tissue is a particular target for noxa, both as part of the affected organ and in its capacity as a body store. There are many different ways of determining presence of heavy metals in tissue.

Any prepared tissue obtained during surgery should not only be tested pathologically/histologically, but also examined for presence of metal. The same goes for tissue obtained from postmortem examinations. Miscarriages and stillbirths are particularly rich sources of interesting findings. Regrettably this type of testing is only rarely ordered, maybe because politics, with the public funds being ever more empty, regards these tests as unwanted.

As to myself, I have been ordering testing of tissue for metals for more than 15 years. At first, my colleagues in the various surgical departments were skeptical, but I gained a lot of help from pathologists. Tissue samples are placed in formalin, as per the usual procedure, and sent to the Bremen medical laboratory. Positive results, i.e., levels of exposure that exceed threshold values in blood and urine, were found in more than 90 % of cases.

By themselves, these findings do not prove health damage. However, if the patient's history, together with the results from clinical examinations and laboratory testing, give grounds to suspect a specific cause for the patient's disease, this result can be an important clue. In the field of environmental medicine, we have to build our mosaic one tiny stone at a time: mathematical reasoning rarely succeeds.

Today, testing tissue sample for harmful chemicals is not yet possible.

Testing dental material and the jawbone
If exposure to heavy metals is shown in dental material, this must be considered an indicator for general exposure. Once metals have infiltrated dental tissue, or even the jawbone, we must assume that they have spread through the whole body. Proof of this can be obtained by examining the dental tis-

sue and with the help of medical imaging, such as computer tomography or, even better, with 3-dimensional techniques (such as 3D CT scanning).

Testing breast milk

According to Bund Magazin (a German environmentalist publication), by now more than 300 harmful substances have been found in breast milk (15.). If a neonate has been exposed to toxins from its first day of life, it should not come as a surprise that it falls ill rather quickly. Nevertheless, nursing mothers are only rarely tested for those substances. Do people simply refuse to believe the existence of these problems?

I recommend women who desire children to test themselves for exposure to harmful substances early enough. The commonly held opinion is that nursing a baby during its first three months is best for both mother and baby. This is certainly true in essence. If anything untoward happens, for example if baby and mother develop complications, they should both be tested for possible exposure to harmful substances. In light of such a suspicion, we should not hesitate to also test the breast milk: it is better to have a terrible end than to have endless terror ...

Regrettably, through the years, there has been a sharp rise in the number of complications that arise in both mother and child. The rapid increase in number of cesarean sections is not just a fad but rather provides an indicator, from a medical point of view, that the female hormone household has been changing. This effect may be caused by harmful substances and/or the on average somewhat older age of pregnant women today. From this it follows that harmful chemicals and metals have had more time to wreak havoc on the organs, which can be clearly seen in the example cases of malignant tumors (see chapter 8.8). The current situation presents us with new and stirring questions, for example as to norm values for female hormones before and during pregnancy for both 20–30 year old and 30–40 year old women.

There must be reasons as to why in Germany every third woman (in some clinics even every second) delivers via caesarian section. In contrast, in Africa 90 % of women deliver their children at home, via natural childbirth. In Germany and other industrialized countries, the question as to the reasons for this is not even subject of discussion. As to these reasons themselves, there are some that we may consider: the fact that women are delivering their babies when they are themselves at an older age, and the increasingly larger height and weight of babies. But this means that we must consider changes in female hormone household that are triggered by exposure to harmful chemicals.

However, since responsible parties do not see, or do not want to see, these correlations at all, no studies are done. To this comes that caesarian sections earn more money.

Analysis of eluate collected from apheresis for presence of harmful substance.
Apheresis is a process that removes harmful metabolites from our body, with the use of a specialized double diaphragm filter. These metabolites form the eluate.

Regrettably, apheresis as a diagnostic and therapeutical tool for dealing with environmental diseases is mostly unknown. In Germany, a regulating committee for dialysis set strict policies as to when, and for what diseases, apheresis may be used. In those cases, the public health insurance companies pay for the costs of treatment.

The possibility of testing the eluate for harmful substances after apheresis is even less known. Probably this type of test was first performed in the Lüdenscheid clinic, about 15 years ago by my friend Dr. med. Richard Straube. At the time, Dr. Straube was head physician of the nephrology department. Today, he is senior physician at INUS medical center in Cham, in the Bavarian Forest. He is a specialist for internal medicine, nephrology, and environmental medicine and has specialized in dialysis and apheresis.

Today, apheresis is a standard method for detoxification treatment in INUS Medical Center. The results obtained from testing the eluate reflect the body's level of exposure to harmful substances. Hence, these results are diagnostically revealing and therapeutically significant. After apheresis, the patient will feel noticeably better. There are three reasons for this.

1. Apheresis removes harmful metabolites from the body.
2. At the same time, depending on the filter size, heavy metals and other harmful substances, mostly bound to protein complexes, are also eliminated.
3. During apheresis, patients can be administered supplementary vitamins.

Since detoxification therapy takes a couple of hours, many different harmful substances are filtered out and eliminated during one single run. In the eluate, we found exposure to noxa at levels 100 to 10,000 times higher than reference values in urine and blood. Heavy metals such as mercury and copper were elevated by a factor of 50 to 100, lead by a factor of 1,000. Levels of harmful chemicals were two to five times the norm value.

These findings prove exposure to the mentioned noxa. We may assume them to be causally responsible for the diseases they correlate with.

How I first encountered apheresis.
My history with apheresis is quite unusual. A close acquaintance of mine was on his deathbed in a neurological ward. Tests that I carried out on him had shown him to be exposed to heavy metals and harmful chemicals. Part of this was due to amalgam fillings that later had been replaced with gold crowns. However, my acquaintance and his wife suspected a company located in the valley below their home to be the main culprit. Time and again, terribly smelling vapours rose to the couple's house and spread through the entire street. Discussions with the manager of the company, with the aim to remedy the situation, had not had any effect.

When, in my capacity as substitute leader of the neurology department, I asked the clinic's head physician for permission to have Dr. Straube conduct apheresis on our patient, he at first refused. I explained to him that our aim was to detoxify the patient and at the same time examine the eluate for presence of possible environmental noxa.

Our conversation lasted about an hour. After that, he got to his feet and said that he had to start his round of patient visitations in the wards. I said to him that I would stay around until he'd give his permission, even through the night if necessary—I had all the necessary things with me to spend the night. He knew me well enough to know that I was serious and eventually he gave his permission to conduct apheresis.

Laboratory analysis of the eluate confirmed my suspicion: the patient showed even higher levels of exposure to harmful chemicals and metals than we had expected on the basis of urinalysis. The measured levels were in part a factor 100 or more higher than norm values for urine. The clinic performed apheresis on our patient two more times, after which she was discharged from hospital. A university clinic was kind enough to continue her treatment. In the following time, apheresis was performed every two weeks, then every four weeks, and eventually only one to two times a year. Today, she goes for a checkup once a year and after 15 years, she is still in good health.

Dental report and material analysis of dental restoration material
In the context of an environmental-medical examination, the attending dentist should always prepare a dental report and conduct analysis of all restoration material used on the patient. If the metals found in the material analysis correspond to those found in urinalysis, this provides further evidence for toxic exposure.

Until ten years ago, amalgam was the most frequently used dental restoration material. Practically every single patient was treated with it. What are some of the supposed advantages of amalgam?

- It is cheap, easily machined, and has a superior shelf life.
- Most of its side effects are chronic and hence hard to prove. (In Germany, patients and their attending physician must present a causal connection between disease and exposure.)

What arguments can we present against the use of amalgam?

- It contains toxic heavy metals such as mercury, copper, silver, and tin, and trace amounts of nickel and palladium.
- Amalgam can be an irritant to the mucous membranes of the oral cavity.
- They are capable of becoming an oral irritant through galvanic action.
- Until 15 to 20 years ago, amalgam fillings were primarily made of copper amalgam, making it highly susceptible to corrosion. Today, we can still observe the after-effects.
- Amalgam has a shelf life of around 10–12 years. These types of fillings are, however, typically worn for a much longer period. In the end, who would have fillings removed that are still within their expiry date and that do not cause any acute problems?
- After fitting, amalgams must be polished, to prevent rapid abrasion. However, out of time constraints or ignorance, hardly any dentists do this.
- Amalgams are, by definition, not a true alloy.

- Consciously or unconsciously, amalgams are not always removed completely. Gold fillings have better grip when they can stick to an amalgam residue basis. Is there any layman who knows this fact?
- Many people wear both amalgam fillings and other alloys. Hence, they reinforce each others' toxic action.
- Dentists are required by law to dispose of amalgam in a special way. It is clear that its toxicity is well-known. Noncompliance is a punishable offence. Nevertheless, amalgam is used day in day out.

Today, we know that so-called precious metal and gold alloys are also dangerous to our health. Essentially all dental restoration materials are foreign substances to our body and may hence lead to health damage. Therefore, people often become a permanent storage depot for whole "batteries of metal", since they carry all these toxins with them when they die. Today, many people opt for cremation after death. This has the consequence that crematories now have to deal with the problem of how to dispose of these toxic substances.

One day I examined a patient who told me that her dentist had removed all her amalgam fillings years ago and replaced them with ceramic crowns. I could not believe my own eyes, when I examined her upper jaw and found 10 amalgam fillings, albeit merely stubs. The patient's ceramic crowns had been placed on top of these amalgam residues. She suffered from asthma and fibromyalgia.

There exist imaging techniques that allow us to recognize exposure to metals in the teeth and the jaws. Amongst them are so-called panorama photos (panorama image of the dental root) and the modern 3D imaging techniques. Panorama photos allow us to determine variations in the teeth, the gums, and the jawbone. This type of imaging also should be included in any environmental-medical investigation. However, it is a precondition for its success that such a procedure is carried out and interpreted by an appropriately experienced dentist. Many dentists cannot recognize, e.g., residues from amalgam fillings that have deposited inside the jaw bone.

The latest 3D-imaging techniques, i.e., a three-dimensional photo, are more modern and even better than the panorama technique. With these techniques it is, in particular, possible to recognize inflammatory focuses in the teeth and the jaw. The disadvantage is that the procedure is quite expensive.

Tests of the immune system

Epicutaneous testing (skin test)

This test serves to determine chronic hypersensitivity to heavy metals, which was in the past designated as "chronic allergy". In fact, this method should be banned for testing for hypersensitivity to heavy metals. Why is this so? Our skin's outer layer consists of horny tissue. A substance to be tested is applied to this partially rather thick layer of skin. However, this layer is not responsible for any hypersensitive reactions, which are in fact triggered by so-called T-lymphocytes.

In an epicutaneous test, a patch that contains the relevant heavy metals is placed on the patient's skin. After four to seven days, they are then examined for any skin response. If any is found, for example reddening or swelling, the result is considered a positive.

Although, according to a leading dermatologist (Prof. Pfeifer, former chairperson of the dermatological society), this test has been obsolete for the last 15 years, health insurance companies continue to require it. If the test is positive, they may partially reimburse the costs for treating amalgam hypersensitivity. Clearly, the insurance companies consciously keep insisting on long-obsolete testing methods in order to keep the number of claims for damages as low as possible.

LTT to test for hypersensitivity to metals—blood testing[30]

A better and hence more meaningful test, compared to epicutaneous testing, is the LTT (lymphocyte transformation test). This test has existed for more than 20 years and is internationally recognized. LTT is based on the immunological principle that specific lymphocytes proliferate after stimulation with an antigen.

The substance to be tested is added to a petri dish with a sample of the patient's blood. The foreign substance then develops antigens against the blood. After its first contact with these foreign antigens, the blood cells' immune reaction creates so-called memory cells, lymphocytes of both type B and type C. In this process, it is the T cells that carry out the specific cellular immune reaction.

---

30 In the past, we used the terms acute and chronic metal allergy.

Advantages of the LTT over epicutaneous testing:

- No allergic reaction is provoked in the patient.
- LTT is more meaningful, because it corresponds to the actual situation inside the human body (our skin, in particular its outer layer, is not identical to our immune system).

9.7.3. Testing for genetic disorders

Today, the conventional medical opinion still holds that many diseases of unknown etiology have genetic causes. Hence, genetic research is flourishing and regards these diseases as their big chance. However, I have a different opinion, based on the results of my own research. Plants, animals, and us humans are all exposed to toxic chemicals on a daily basis, which we absorb through respiration, the skin, our nutrition, or medical drugs. Hence, we can observe that, similar to us humans, animals are suffering from increasingly more diseases. In fact, people who wear dental restoration materials are even exposed to metals day and night. These metals get metabolized by our body's own enzymes and can then be eliminated.

These enzymes include the following substances:

- N-acetyltransferase 2 (NAT2)
- Cytochrome P450 (CYP2C9)
- Cytochrome P 340 (CYP2C19)
- Glutathione-S-Transferases M1 (GSTM1)
- Glutathione-S-Transferase P1 (GSTP1)
- Glutathione-S-Transferase T1 (GSTT1)

If there exists a genetic defect in the synthesis of these enzymes, harmful substances and medical drugs cannot be sufficiently metabolized by and eliminated from our body. More than 70 % of cancer patients carry such a genetic defect. This data coincides with what can be found in the scientific literature—these are worrying numbers.

Once again, it is primarily metals that cause these genetic defects. Hence, I regard them as a consequence of poisoning with harmful substances. It is possible to prove this enzymatic damage, which can even be partially repaired after removing their cause (the metals). Hence, we are not just dealing with a new diagnostic principle, but also with a new therapeutic idea that targets the cause of the disease—this in contrast to chemotherapy and radiation treatment, which strictly fight the symptoms.

9.7.4. Imaging as diagnostic method

Brain-SPECT and PET

Brain scanning with SPECT and/or PET (positron emission tomography) is indicated in particular if there is reason to suspect damage to the central nervous system by toxic substances. These techniques are capable of probing the metabolic functioning of the brain. These procedures should always be preceded by an MRT scan (magnetic resonance tomography). If the MRT scan is negative and SPECT or PET scans are positive, this points towards a cerebral metabolic disorder, caused by harmful substances. The foci of the disorder are visible even to the layman if one explains the scans to them. Here too, it is a precondition that the examination is conducted by a specialist in the field. Many radiologists have not yet mastered these techniques and the interpretation of the results.

9.7.5. Exposure to harmful chemicals and fungi toxin

The dangers posed by chemicals lurk in every corner. In the case of harmful chemicals and fungi toxins the diagnostic procedure is similar to that for suspected heavy metal exposure.

In all cases, we first search for evidence of these substances in the blood. Since different test tubes should be used for each different substance, one should first contact the laboratory, which will then send the tubes together with an instruction sheet.

In the most severely diseased patients, apheresis should be performed after finding exposure to harmful chemicals, and the eluate examined in the laboratory. One is often surprised at the high concentrations of noxa that are found. Often, patients feel noticeably better already after the first apheresis—a fact that also points toward a causal connection between disease and exposure to harmful substances. For completeness, I would like to mention some further environmental-medical diagnostic tests, namely the so-called biological effect monitoring and so-called susceptibility monitoring. These tests should only be conducted by trained specialists.

If fungal infestation is suspected, the first thing to do is to consult a professional to survey the rooms in which the patient typically resides (house, work). As a principle, I surveyed these spaces in all cases of patients in whom I expected exposure to fungal moulds. If I noticed any hot spots of humidity with signs of fungal infestation, I ordered humidity measurements in the interior rooms (along the walls, ceilings, and floors), or conducted these measurements myself. If these tests were positive, I proceeded with surveying the rooms for fungal infestation. This type of survey is done by, e.g., institutes for interior diagnostics. In my case, I cooperated with the Dr. Lorenz institute in Düsseldorf.

If the results of these surveys were positive, I continued diagnostics by testing the patient's stool for fungi. It is also possible to test for sensitization to fungi, with the help of LTT. If the results from these tests overlap with those from the interior measurements, we have obtained proof of a causal connection between noxa and disease that is also sufficient in a legal sense. In all other cases, if either only the patient is affected or if only the room surveys were positive, a medical connection is highly likely—and hence we should proceed with treatment—but this connection is only rarely legally recognized. One should be aware of this fact, if one wants to sue their landlord.

9.8. Alternative medicine—drugs and treatments without side effects?

9.8.1. Dental organogram

A dental organogram tries to answer the question as to the relation between diseased teeth and diseased organs. Like other parts of our body, teeth also developed from the three germ cell layers entoderm, mesoderm, and ectoderm. In his book "Biophotons—the Light Inside Our Cells", M. Bischof (10.) points out that Carl Hutter and Rudolf Steiner already realized that tissue and organs that developed from these three germ cell layers are in fact organ systems, i.e., functional units. The interplay between these units are an additional factor influencing our physical and mental condition and determine our individual type of constitution.

Diagram showing organs and body parts that are at risk due to toxic foci inside of/near to/underneath of the teeth and jaw. This diagram is 3,000 years old! A specific toxic focus also slowly poisons adjacent areas.

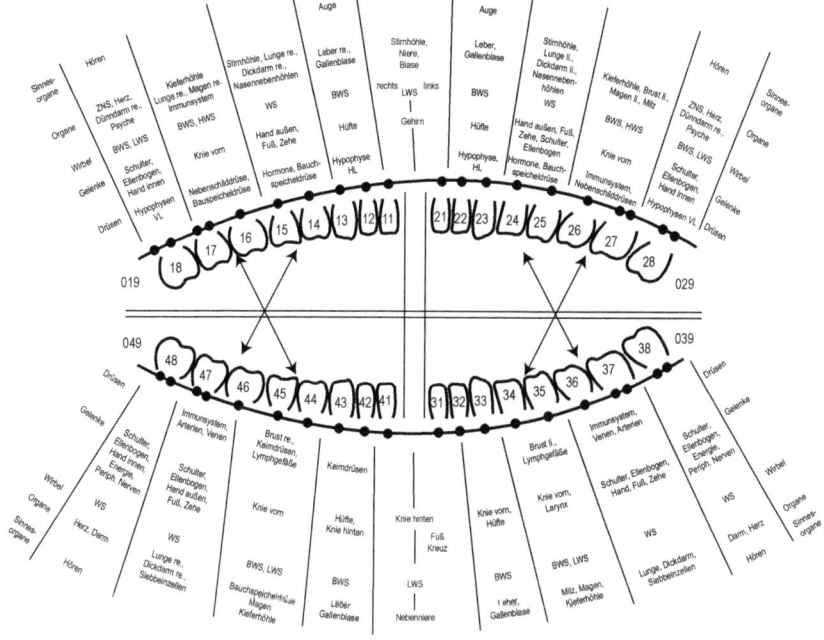

Zahnogramm
So „lesen" Sie Ihr Röntgenbild, so spricht Ihr Zahnarzt von Ihren Zähnen.
Bleibendes Gebiß:
Benennung von 1 bis 8; zur Festlegung der Seite und ob oben oder unten liegend wird noch eine 1, 2, 3 oder 4 davorgeschrieben

rechts oben								links oben							
18	17	16	15	14	13	12	11	21	22	23	24	25	26	27	28
48	47	46	45	44	43	42	41	31	32	33	34	35	36	37	38
rechts unten								links unten							

47 = vier-sieben = rechts unten der 7. Zahn, der 2. Mahlzahn | 12 = ein-zwei = rechts oben der 2. Schneidezahn

Typical signs that indicate a toxic focus are: a dead tooth, amalgam deep inside of the tooth (near the dental root), amalgam underneath gold, amalgam splinters in the jawbone or underneath the dental root, but also the presence of bacteria and toxins that have become embedded in toothless parts of the jaw. The latter is the most frequent cause of a chronic zinc deficiency, rheumatism, and cardiac problems. Unilateral dental foci lead to unilateral brain damage that causes physical weaknesses on the opposite side.

Figure 28: dental organogram

The above was and still is well-known in China, as I learned during my studying of traditional chinese medicine (TCM). In TCM, the interplay between organs and organ systems plays a large role—something which one gets to know intimately as a candidate for the exams: if you are not capable of thinking in the same terms as TCM, you'll fail the exam. It was mandatory to shift your mental context towards the Chinese way of thinking and life philosophy.

Due to my experience with applying this way of thought to my placement of acupuncture needles in my own practice, it was not difficult for me to grasp that if a tooth (or part of it) is diseased, organs that developed from the same germ cell layers can also be affected. The connection between teeth and organs can be seen in the dental organogram.

In her book "Dangers Posed by Dental Restoration Materials and Environmental Noxa", Theresia Altrock (1.), a gynaecologist with years of practical experience, described toxicological-pathogenetic correspondences, mostly with the help of resonance methods. She showed the relation between dentistry and gynaecology. The, as the foreword to her book formulates it, "culmination of my work" is her confirmation and application to her daily practice of the second hermetic principle—in short: "as above, so below, and vice versa"—on an anatomical-functional level.

My own analysis of the dental organograms obtained from more than 100 patients has shown that this interconnectedness exists in all participants. Hence, we may regard the dental organogram as a further important diagnostic tool.

Hence, I can recommend all my colleagues to also apply this diagnostic method. It is a real asset to our toolbox and provides more diagnostic certainty, without the use of mechanical devices. Another advantage is that it does not cost a single cent.

Acupuncture

I have been occupying myself with acupuncture, homeopathy, and naturopathy—that is, medicine without side effects—since 1981. As a student of F. Bahr (3.), director of the German and European academy for acupuncture and side-effect-free medicine, I also familiarized myself with RAC (reflex auriculo cardiac) after Nogier (91., 92.) and came to value the techniques that it teaches.[31]

Diagnosing disturbance foci with the help of acupuncture

Disturbance foci are noxa and foci such as diseased teeth, scars left behind by operations or injuries, chronically inflamed and partially scarred tonsils, etc. In the diagnostics of these foci, there are five levels of severity.

In order to help understand this, one should imagine our body and its immune system as a water purification plant consisting of five cascading basins. If the first basin (first immune system) is full, the remaining toxins overflow into the second one, etc.

When using this disturbance foci technique, it is easily possible for an experienced practitioner to determine the level of severity of the disturbing factor. If there exist multiple foci, they can be classified according to the level of danger they pose. During treatment, it is possible to track the patient's improvements by conducting checkup tests on the foci. In the hands of an experienced practitioner, this is a very elegant method.

> *Nogier and Bahr must be credited with introducing ear acupuncture (auricular medicine). They found five points in the ear that they assigned to disturbance foci. With RAC (reflex auriculo cardiac = ear-heart-reflex) it is possible to test any substance for tolerability.*

Many conventional physicians will shake their heads in disapproval. This does however not stop me from bringing this method to people's attention. I am astonished that only few colleagues are prepared to learn alternative methods.

Disturbance foci diagnostics is ideal as a screening method and to check the progress of the healing process. It has, however, one disadvantage: it is not measurable—neither quantitatively nor qualitatively—leaving patients

---

[31] Reflex auriculo cardiac (RAC) was discovered by the French physician Dr. Paul Nogier, in 1966. RAC-testing is a diagnostic method that has its roots in ear acupuncture (auricular acupuncture). By taking the patient's pulse in a particular way, the correct acupuncture points can be found. Additionally, allergies, toxic exposures, and drug tolerability can be tested. The test has no side-effects, costs hardly a single cent, and can be conducted within a few minutes, in any medical practice, to test, e.g., DMPS. Nevertheless, one must have received special training and invest a large amount of time in gaining practical experience.

and doctors no data to show to third parties. In case of failure or in the context of legal procedures, this can be a disadvantage for the patients concerned. However, it offers an additional clue: if the patient improves after the disturbance focus has been removed, this can be exploited as evidence, in particular when there are multiple success stories to help corroborate the argument.

We are still unilaterally imprinted by the scientific-mathematical point of view. For medical and legal reasons, diagnostic and therapeutic findings must always be measurable. These findings are also important for monitoring the patient's progress. If laboratory results proceed in step with a deterioration of or improvement in the patient's condition, this points toward a causal connection.

In China, all throughout Asia, as well as in the USA, acupuncture is used for therapeutic purposes on a daily basis. Why shouldn't this be possible in Europe—in Germany?

9.9. Experience and its implications for diagnostics

Many people will ask: what does experience have to do with diagnostics? My answer would be: a lot. Experience is of use when investigating causes and sources of in particular environmental diseases, because quite frequently these investigations require detective skills. In his book "Intelligent Cells", B. Lipton (73.) describes how experience controls our genes. Genes can be inherited. They can be damaged by harmful substances, but also by the psyche, and now we learn that they can even be controlled—this is even proven by science.

It would be beyond the scope of this book to discuss these connections in further detail. It is, however, a matter of concern to me to at least mention them, because these are new approaches to thinking about the subject, which have to some extent already been scientifically proven. Lipton's book is simply fascinating!

Bottom line
Our mind and psyche can affect matter, something which is already reflected in psychosomatic medicine. To me, it is a new and exciting discovery that they are concretely capable of controlling our genes. How is this supposed to work?

Lipton's insights help us better understand F. A. Popp's discovery of biophotons, of which M. Bischof (10.) reports in his book "Biophotons—the Light Inside Our Cells". Bischof writes that the Russian physician Professor Alexander Gurwitsch was the first to discover light inside the cells of onion roots.

After Popp, many researchers from all over the world have confirmed that cells from all living species emit light. Popp was able to show that we are dealing with coherent light[32] that, similar to laser light, possesses a high degree of order and is hence optimally suited for transmitting information. Clearly, this light plays an important role in the interplay of the various organs and hence in our allover physical and mental constitution. Here then, we come full circle from our starting point in chapter 6.5, where I mentioned electromagnetic radiation and coherency.

It can only be hoped that conventional medicine starts applying these groundbreaking insights into life and the functioning of the human body as soon as possible. Many naturopaths already do this. As an aside, it is a hardly-known fact that the number of new private naturopathy practices has dramatically increased, for example in Mecklenburg/Vorpommern, Brandenburg, and in the major cities.

---

32 Coherency (from the latin root cohaerere = to be connected) is a concept from physics that designates waves' capability to act according to shared rules during their dynamic development.

10. Therapeutic options for treating environmental diseases

10.1. Introduction: so you are suffering from an environmental disease. What now?

So you are suffering from an environmental disease. What now? What do you do when your prior history has led you to possibly—or even with high probability—be a victim of environmental toxins, i.e., a victim of an environmental disease.

In these cases, provided that conventional medicine has not been able to help, one should consult an environmental physician. The whole field of environmental medicine, however, would remain stuck in theories and advices if it were not able to find treatment options as an alternative to conventional medicine—or, even better, in conjunction with each other. Finding these options should be the target and, following from that, the foundational basis of any environmental-medical discourse. For this reason, I would like to sketch these new therapeutic options in somewhat more detail.

In the process, I will once more limit myself strictly to the approaches that I use myself.

> *Finding therapeutical treatment fitted to remedy the underlying causes of disease must be the target of any environmental-medical action. Viewed this way, environmental medicine is both cure and prevention at the same time.*

Seen in the light of the above statement, environmental medicine is radically different from conventional medicine as it is currently practiced. This kind of lasting remedy is now possible, thanks to new, scientifically founded therapeutic options. They are logical and intelligible for the layman, in particular the affected patients and they could, nay, should open up new avenues for treatment to colleagues in the medical profession.

This is necessary, since the number of physicians with knowledge about long-term, comprehensive environmental-medical therapies is very small. The information presented in the following can be a valuable addition to the knowledge of anyone working in healthcare.

Cooperation between a patient's family physician and an environmental physician is a precondition for successful treatment. If the disease was caused by dental restoration material, a dentist also has to be involved, who should have some experience in the field of environmental health. Curiously, in contrast to the general cluelessness amongst the general medical profession, the number of dentists with environmental-medical experience has been steadily increasing. Apparently, our colleagues in dentistry are becoming increasingly aware of their responsibilities: we conventional physicians make our lives more difficult, in that respect.

Any treatment should be tuned to each individual patient and be conducted in a most considerate way, i.e., any medication used should be well-tolerated by the patient. It is advisable to run pharmacogenetic and immunological test on any drugs with many side-effects, to find out if they are metabolized by the human body and if they are tolerated well. In the tumor study described in chapter 8.8 I determined levels of glutathione-s-transferases, which are essential to the metabolism, and hence the excretion of harmful substances. An LTT can be used to test if the patient tolerates the foreign substances (chemicals, medical drugs).

Any therapeutic approach must be tuned to the kind of diagnosed substance exposure. If the patient was exposed to multiple harmful substances, treatments can progress in parallel.

It is important to fully inform the patient about the nature and length of any therapy before commencing. Even though there are numerous cases where rapid healing could be enacted, most frequently treatment can be expected to need several months or years, which the patient should be made aware of in advance.

That which has been caused by many years of toxic exposure cannot be cured in a few weeks.

Additionally, patients must be informed that temporary improvements in their situation can be followed by a relapse. Hence, regular checkups and, if necessary, further treatment regimens might be expected.

Like conventional medicine, environmental medicine is not capable of healing everything. The patients must be informed that protracted exposure may have led to lasting damages. This calls for honesty from the treating physician but creates a situation of trust. If one promises much but isn't capable of keeping those promises, one only creates displeasure or fear.

There exist therapeutic strategies other than the ones I describe: I only mention medication and treatment options that I have tried in my medical practice and which I can hence recommend.

I differentiate between treatment options for

1. heavy metals
2. harmful chemicals
3. fungal toxins.

10.2. Detoxification with the help of drugs

10.2.1. Detoxification after heavy metal exposure

We must differentiate between heavy metal exposure (and no exposure to other substances) caused by dental restoration material, or by professional or private exposure to, e.g., exhaust fumes or skin contact. In the search for causes, we also have to consider domestic sources such as tap water, should old plumbing made of lead or rusty copper be present in the patient's home.

In the case of heavy metal exposure, we should keep in mind that in the human body metals have a half-life[33] of 15 to 20 years, which means that without appropriate detoxification these substances are never completely excreted.

Chemical detoxification with DMPS

Detoxification may be initiated by administering chemical compounds, such as DMPS. In a later stadium, a naturopathic basis is appropriate. Children and severely exposed patients often have little tolerance for "chemistry". In these cases, we must opt for very low doses and proceed carefully by at first intravenously injecting only this small dose, which can than be slowly raised. I often draw a little blood, in order to dilute the injection, or administer the medications via a slow drip. I never encountered any complications whilst using this procedure. In case of dubious tolerability of a particular medication, I used acupuncture for additional testing.

---

33 The time period needed to metabolize 50 % of a substance.

Our agent of choice is DMPS (2,3-Dimercapto-1-propanesulfonic acid), administered as an injection, or Dimaval, given in capsule form. This compound is a chelating agent and forms a water-soluble complex with copper, mercury, lead, tin, iron, cadmium, nickel, and chrome. The degree of exposure plays a role here, since in the case of for example high levels of exposure to copper, only a relatively low amount of mercury is excreted.

Mechanism of action of chelating agents
One characteristic of chelating agents is their pinch-like function (101.). The word 'chelate' stems from the Greek word for lobster; chelating agents, much like a lobster's pinchers, grip their target and bind with heavy metals to build cyclic complexes.

Chelating agents have the advantage of being capable of diagnosing both acute and chronic metal poisoning and allow for checking the success of a detoxification regimen. They remove toxic substance from the human organism and prevent the development of numerous chronic diseases. Hence, they have use in diagnostic, therapeutic, and preventive medicine. In contrast to naturopathic agents, chelating agents have a rapid effect and, if used appropriately, rarely have any side effects. They can be administered by any medical physician.

In the case of acute diseases (such as cardiovascular diseases or infection), treatment with Dimaval (DMPS) is not indicated: treating the acute disease always comes first.

> *We should strive to introduce detoxification therapy in every medical practice and in every single hospital.*

In as little as a few weeks, both the patient and their attending physician will recognize the advantage of chelating agents and appreciate their merit.

Public health insurances will be amazed at the rapid decrease in money they have to pay—exactly as I encountered in my own practice: when I started out, the costs of the medications that I prescribed was 15 to 25 % above average. I always wanted only "the best" for my patients. However, ever since I've started working as an environmental health physician, the costs I declared were 15 to 20 % *below* average.

There are patients who literally "cry" for an injection. I encountered this in particular in MS (multiple sclerosis) patients or patients with other neurotoxic diseases. In all three cases I found exposure to metals. They managed through four weeks without any injections but then intensively craved their next shot. One should know that injecting DMPS in MS patients is officially contraindicated—as students of pharmacology we were drilled to believe that DMPS may only be injected by a medical specialist and only in cases of acute poisoning.

Hence, we arrive at the question of how to recognize exposure to metals in patients. This can only be done by giving DMPS. Hence, I recommend to inform each patient of the risks, before administering the first injection, in particular in the case of MS patients. Ideally, the physician should obtain written informed consent from the patient. When following this protocol, I never encountered any complications: MS patients thanked me for each single injection because, as they told me, it gave them a few days or even weeks of pain relief and gave them better mobility.

DMPS: therapeutic regimen
One DMPS injection per week, or every two to four weeks, or one Dimaval capsule every third day. One can replace Dimaval with one capsule of DMSA, which has preference if the central nervous system has been affected, because DMSA can pass the blood-brain barrier.

DMSA: therapeutic regimen
First series (14 day): DMSA 100 (or 200) divided over three days; vitamins E and C and supplementary zinc (e.g., zinc orotate 20 mg or 40 mg) during the next eleven days.

This first series is followed by two further series. One week before initiating detoxification treatment, I give the patient hepatica and solidago drops, to support liver and kidneys.

Why do I administer vitamins and zinc?
Vitamins E and C further promote detoxification: I administer vitamins B12 and B6 if I suspect poisoning. In the case of B12 I prefer intravenous injection because it works more rapidly and effectively. It can, however, also be injected intramuscularly. I usually administer vitamin B6 in capsule form—every time, one is surprised again at how this speeds up recovery.

I first encountered the need for vitamins and growth factors during the recovery process when I was in Africa. Growth factors are a concept from cell biology by which we mean specific proteins. They mediate information in the form of signals between two cells and through this mechanism stimulate cell growth by binding to a specific receptor located on the cell membrane—government-run hospitals in Africa were known for treating their patients almost solely with yeast tablets: there was no money for medical drugs. I knew about the reasons for these tablets' effectiveness because I wrote my Ph.D. thesis on the extraction of cell-multiplication-stimulating substances from yeast extracts (127.). Yeast cells contain vitamins (e.g., vitamin H), nucleosides, and nucleotides, which are building blocks of our DNA (deoxyribonucleic acid) that stimulate cell growth, and hence cell multiplication.

In addition to the vitamins, I administer zinc compounds. Zinc is an antidote to other heavy metals: when administered in excess, it slowly displaces other heavy metals from the body's stores.

Since many metals are excreted via the kidneys, it is important to consume lots of fluids during the detoxification period, ideally three to four liters per day. In Germany, the average person drinks ca. two liters of fluid.

I frequently started treatment by giving Dimaval and continued with DMSA if I found any affliction of the CNS (central nervous system), after which I changed to a naturopathic treatment (with microalgae).

10.2.2. Detoxification for industrial exposure

The most important therapeutic principle is to find and avoid the source of exposure: rusty lead or copper-containing plumbing should be removed and the workplace should have sufficient ventilation. Additionally, every room must be frequently aired. In my two and a half year's time as occupational health physician, I was surprised time and again at how rapidly companies took measures to reduce harmful substances in the workplace once one or more employees developed an occupational disease. Why hadn't the toxic risk not been removed before? Why haven't employees been quicker to notify their employer? The answer: in most cases, they were afraid to lose their jobs.

In the case of exposure to chemicals (PCB, formaldehyde, wood preservative, solvents) we must also consider the patient's own residence as possible source. In these cases, the interior of the patient's house must be examined by an expert on interior diagnostics (for an address, see the appendix).

Some of the tougher cases necessitated using tracking dogs, the cost of which was carried by the health insurances after its necessity had been sufficiently proven. It was impressive to watch those dogs doing their job. Before consulting an interior diagnostics expert, I inspected the house in order to convince myself of the necessity of further diagnostics. In addition, it was my ambition to find the source of exposure before consulting an expert—something which can be done once one gains sufficient experience and practice.

Detoxification proceeds along the same lines as outlined above. Like heavy metals, harmful chemicals collect in the human body's stores, in particular adipose tissue, liver, kidneys, and the brain. In contrast to metals, however, there is no specific substance to help eliminate these chemicals from the body. In these cases, immediate avoidance of the source is of utmost importance. From a therapeutic point of view, we need to strengthen the immune system, e.g., by administering vitamins.

If one follows this procedure, the results are more than just a recompense for patients who cannot be sufficiently helped or not helped at all. In any case, the patient's health will not worsen any further. This already means a lot, since patients are used to their health deteriorating despite all medical and therapeutic efforts.

10.2.3. Dental restoration and detoxification

If health damage due to amalgame is proven, I recommend getting an expert report from both the patient's dentist and their attending physician to prove immediate necessity of dental restoration work, in order to get the patient's health insurance to carry the costs of treatment. The last is even more difficult if additional exposure to other metallic compounds has been shown, since the health insurance companies—with rare few exceptions—refuse to pay for restoration work, even if a causal connection between metals and diseases can be proven. Nevertheless, one should always keep trying.

As described in chapter 7.2.1, we frequently encounter cases where amalgame has been removed and new dental restoration material has been applied without prior detoxification. The consequences of it all are then even more aggravating, and detoxification becomes even more laborious. If exposure to heavy metals caused by amalgam has been proven, one should first undertake dental restoration and then start detoxification treatment. In order for this to work, both medical professionals and dentists must cooperate closely, which is beneficial for all three parties involved—the patients, the dentist, and the family physician.

The mouth, with its teeth, is the "portal to our lives". If there are any diseases in this "foyer", all of our nutrition and everything that happens inside the oral cavity can become a source of harm to the human body. In this respect our teeth and all dental material have a high medical significance. J. Lechner (71.) wrote his book "Healthy Teeth—Healthy Human" with this in mind.

We may reproach the parties responsible for managing the medical and dental associations for not sufficiently cooperating in the past. Dentists have too little medical training and use (even up to perfection) any restoration material that industry offered them without any thought for future effects.

> *We as medical physicians have laid too little significance on the mouth as a portal to the gastrointestinal tract and simply did not pay enough attention to dental restoration materials as substances foreign to the human body. Once the discussion finally started, it solely revolved around mercury and amalgam. Any danger to our health posed by other metals and materials has effectively never been considered.*

From 1926 onwards, when amalgam was first banned in Germany, dentists, medical physicians, and manufacturers of restoration materials should have worked together more intensively. In the USA, manufacturing and distribution of amalgam was first prohibited as early as 1839. Hence, the dangers posed by dental restoration materials have been known for more than 150 years.

> *From this all it follows that cooperation between dentists, physicians, and manufacturers is urgently needed.*

During my research over the past 25 years, I have not encountered any such studies conducted by manufacturers of dental restoration materials. I have contacted these companies and urged them to conduct long-term studies on dental materials. All of them refused. As reasons I was given that it might cause trouble and we must safeguard jobs. No further details were provided.

At least after the amalgam trial, the respective studies should have been brought forward as testimony and proof. At this point, the reader is once more encouraged to stop and think about the question as to why these clarifying studies have still not been done until now. They are unavoidably indicated from a medical but also from moral-ethical point of view and could have been done quite easily with the help of the diagnostic and therapeutic methods that have become available over the last 15 years. The Frankfurt amalgam trial involved 1,500 patients with environmental health issues. In 1995, the number of patients with amalgam and dental-restoration-material-related damage alone was as much as 400. We as medical physicians and dentists must ask ourselves if we failed at clarifying the amalgam problem.

Back to dental restoration. I postulate ten rules, which I agreed upon with the patient and the attending dentist before commencing treatment.

Ten rules for dental restoration work

1. The patient must feel in good health.
2. No dental alloys should be removed at any time during pregnancy.
3. Detoxifying medication must be taken before dental restoration material is removed, e.g., one capsule of Dimaval two hours before treatment.
4. Material must be removed quadrant by quadrant, protected by a dental dam, with influx of oxygen.
5. The dentist must use a low-rpm drill which drills around the amalgam in close proximity to the material to be removed.
6. Not more than three or four fillings should be removed per week.
7. High-quality cement (e.g., glass ionomer cement) must be used for a temporary filling.
8. After removing all fillings, therapy continues with the actual detoxification.
9. After detoxification, the patient's reaction to the new material must be monitored until the DMPS test becomes negative.
10. Finally, the new, permanent filling material is introduced.

10.2.4. Apheresis: only in the most severe cases

In severe cases of proven exposure to metals, chemicals, solvents, etc., especially if the patient's life is in danger, apheresis is the best treatment. In the case of tumors, this procedure is called oncopheresis. Since this procedure is largely unknown, I will outline it in somewhat more detail.

Apheresis was developed in Japan as membrane-differential-filtration (MDF) and was the first method using lipid apheresis to circumvent the disadvantages of plasma exchange. Therapeutic apheresis (TA) is a particular form of high-tech plasma purification and immunomodulation to treat chronic and acute metabolic and immune-system disorders – so-called auto-immune disorders.

With the help of blood plasma purification, also called plasmapheresis (from the greek root 'apherein' = to separate), harmful proteins and metabolites (e.g., cholesterol) or toxic substances are removed from the human body. Among the removed substances there are what we call, for ease of speaking, protein bodies that can bind to tumors (tumor-associated proteins), circulating immune complexes, and complex inflammatory toxins—i.e., including metals and harmful chemicals, which are transported by binding to precisely these protein bodies. This type of detoxification treatment restores the human body's natural equilibrium.

During apheresis, no plasma exchange takes place and hence hardly any electrolytes or antibodies are lost, making it a very gentle procedure (114.). It is in use for treating exposure to environmental toxins, rheumatic diseases, disruption of lipid metabolism, chronic inflammations, secondary diseases that follow from infections (such as borreliosis) as well as autoimmune diseases.

Lipid apheresis removes certain fatty acids from the body. Upstream membrane filters remove plasma components with high molecular weight. Hence, any toxins bound to high-molecular-weight protein bodies can be removed from the blood plasma (63.).

Apheresis is capable of removing harmful substances of any kind.

Disadvantages of apheresis, compared to other detoxification strategies

- Apheresis is disproportionately more expensive than chelation therapy.
- Substances are only removed from the blood and not from the body's stores. Hence, apheresis should only be used in severely ill patients, where it can often save lives.
- The argument that apheresis can save lives must most definitely be expressed in any report that is to be submitted to the health insurance company.

10.2.5. Treatment of fungal toxin poisoning

We now arrive at a topic that is becoming increasingly more important, namely diseases, specifically poisonings, due to fungal toxins. Here too, the first therapeutic step is to avoid and remove the source of infection. Specialists must dry out humid walls, roofs, and houses as a whole, so that the basis of their proliferation is removed.

As first therapeutic step, the affected patient must make changes to their lifestyle, amongst which count vitamins and physical exercise. One should be careful in using antifungal agents (antimycotics) especially if the bowel is affected and in particular when a comorbid exposure to metals is present. The reason, unknown to most, is that although antimycotics kill the fungi—and hence lead to temporary improvement—this process releases heavy metals from the body's stores, which can then attack the organism and invade the organs' stores. This is because fungi can bind to heavy metals—they are capable of almost literally devouring them. When this happens, inorganic metals are changed into organic compounds in a process called biomethylation[34].

However, this step does not end the process of assimilation in the body. On the contrary, it is only the beginning: metallic compounds can be mutually interchanged via transmethylation or transbutylation. For example, dimethyltin can be converted to trimethyltin and dibutyltin to the even more toxic tributyltin (TBT).

What does this mean in the context of treatment with antimycotics? Killing off the fungi leads to a release of highly toxic organometallic compounds, which infiltrate the organ stores. Like mercury, organic tin compounds show a particular affinity for the central nervous system (CNS). They cross the blood-brain barrier and bind to ganglia (the nervous systems' switching boards), nerve cells, and nerve fibres, thus disrupting their function. Hence, CNS disorders are not only typical for exposure to mercury but are also increasingly suspected consequences of damage by organic tin or mercury compounds.

---

34 Biomethylation is the binding of one or more methyl groups (-CH_3) to some metals or metalloids, mediated by microorganisms. These transformations create organometallic compounds.

The examples of our female patient with the brain tumor (glioblastoma) and of Matthias L. show that with high likelihood all metals are capable of crossing the blood-brain barrier, in particular if the patient's immune system is weakened.

> *From this it follows that one should first detoxify a patient of metals before removing the basis of fungal proliferation. Only after this has been done we should determine the further course of action.*

Fungi are, in fact, even capable of changing one metal into another. We still do not know how this process works, but when we observe the periodic table, the differences between metals, for example mercury and lead, are not very large. By administering antibiotics, we possibly cultivate new, even more resistant, fungal strains, which then, in their search for new nutrition, so to speak, assimilate new metallic compounds.

In a way, they do the human body a favour: by saying "We live in symbiosis with metal-retaining fungi. We need each other." With respect to metals, our body answers, figuratively speaking, "It's better for you guys to be eaten by fungi than to infiltrate and damage the organs." The human body exploits this symbiosis to maintain the equilibrium; an equilibrium that also is the most important therapeutic target when treating environmental patients—removing the cause is the first precondition. Treating the symptoms alone will only lead to new complications, new diseases, and further disturbance of the equilibrium. Hence, treatment of fungal infections should always be in the hands of a specialist physician.

10.3. Detoxification with side-effect-free medication— naturopathy

I can only recommend all of my colleagues to familiarize themselves with the principles of naturopathy. In no way should conventional medicine suffer from this. On the contrary:

> *Naturopathic treatment is an excellent complement to conventional medicine, including environmental medicine.*

Traditional Chinese Medicine (TCM), as practiced in China for hundreds, even thousands of years, should also find its place in our Western world.

Here I once more, explicitly only discuss naturopathic means with which I have experience and which I prescribed to my own patients. Some of these are rather unknown even among registered naturopaths. However, these medications are excellent for detoxification treatment and are free from side-effects. This group of medications contains for examples the zeolites (healing stones), which are almost completely unknown in the West, despite them having been in use for thousands of years.

10.3.1. Healing stones

Stones that are capable of healing. To many colleagues this sounds like madness—how is this supposed to work? Stones are inert materials, aren't they? However, exactly this last point is a misassessment. M. Gienger's "Lexicon of Healing Stones" (41.) says: "Each mineral, i.e., each element contained in a mineral, has properties specific to the mineralogical as well as the healing qualities displayed by a healing stone."

These healing properties have been put to the test many times. In order to understand, one must understand the composition of such stones. They consist of metals and nonmetals. Gienger continues: "The essential properties of whole classes of minerals are determined by nonmetals, such as sulphur, fluorine, chlorine, oxygen, carbon, phosphorus, and the metalloid silicon (in the form of silicate); metals (such as sodium, potassium, magnesium, aluminium, copper, iron, manganese, and many others) on the other hand determine the specific properties of a particular mineral. Metals and nonmetals thus act as a mineral's first name and surname, respectively. The first name (metal) is characteristic of the individual mineral and the surname its family relationship."

What is it that is so special about zeolites? What is the basis of their detoxifying properties? Zeolites (silicon dioxide) belong in the group of aluminium silicate minerals. Silicon dioxide is one of the oldest health remedies known to mankind; it occurs, for example, in clay and has been used for skin care and wound healing for thousands of years. According to R. Straube and K. Hecht (114., 50.), zeolites have as a particular property their capability of forming so-called nano- and Ångström lattices of most diverse shapes. For this reason, they are semimetals, hermaphrodites so to speak, with both metallic and nonmetallic properties. Maybe this is what makes them so interesting and meaningful.

Zeolites occur in three states: phase-pure, flaky, and crystalline. Due to the possible importance of zeolites in the future, I will now discuss them in some more detail.

Clinoptilolite is a zeolite from the crystalline group and is the only one that can be used in humans and animals. According to the description by K. Hecht, it is a microporous tuff with a channel-forming crystal lattice. The interstices have a width of 4 Ångström (1 Ångström = 0.1 nm). In the crystal lattice's channels and interstices we find cations such as calcium, magnesium, sodium, and potassium, partially in conjunction with water of crystallization (non-free H_2O). These cations can easily participate in ion exchange. I simply find these descriptions both exciting and convincing, a sign of the stone possessing motility, yes, even life. It immediately fills my "biochemical heart" with delight. Practicing environmental medicine is an utterly fascinating occupation.

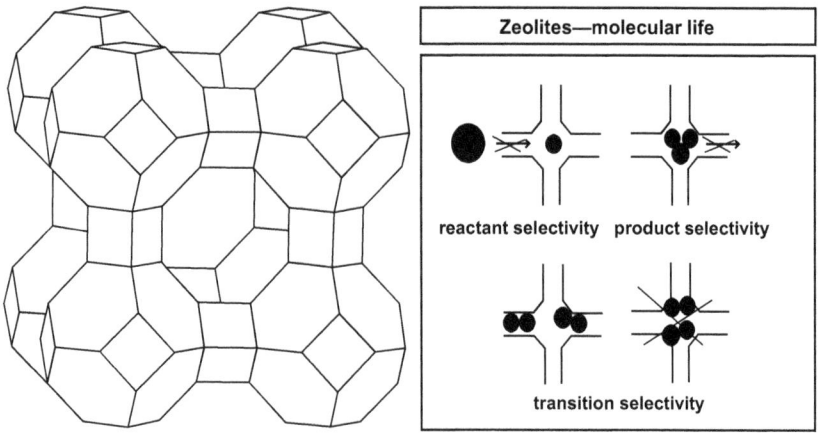

Figure 29: Three-dimensional zeolitic structure

Some capabilities of Clinoptilolite

- extensive supply of minerals
- supply of silicon dioxide
- systemic regulation of the body's mineral balance

- charging the biologic battery, in order to ensure optimal human bio-electricity (EEG, ECG, EMG)
- detoxification and purification
- capturing free radicals
- strengthening the body's capacity for healing itself (including the immune system)
- clinoptilolite is an antibacterial, antiviral, and antimycotic agent
- zeolites can eliminate radioactive radon from the air. Hence, they can be used as a preventive measure in the fight against primary bronchial carcinoma caused by radon dust[35]—the second most common cause of bronchial carcinoma worldwide. (113.)

From this abundance of properties we can easily see how useful this type of zeolite can be to humans. I myself take a regular, yearly zeolite regimen with, e.g., Toxaprevent made by Froximun AG.

Further beneficent zeolite properties:

- no side-effects
- no risk of overdose
- can be used by children and pregnant women
- helps slow down the biological aging process in the elderly
- no risk of developing tolerance or addiction

In 2001, a cooperation between the Catholic University of Freiburg (KH), the Hongkong University of Science and Technology, and R. Schraube accidentally discovered a new zeolite property: elimination of harmful substances and fine dust from the interior atmosphere. The SBS (Sick Building Syndrome = diseases caused by toxic interior gases and circulating particles) is one of the diseases that can be triggered by exposure to interior air that contains toxic substances. Studies have shown that SBS can be treated by removing the toxins from the air. This finding has been applied to curing other diseases and new zeolite-based drugs are in development, such as Toxaprevent.

---

[35] Fine-grained radon dust is a mixture of minute dust particles containing radioactive substances that can infiltrate the alveoli. Similar to black lungs in the case of miners, this dust slowly causes chronic bronchitis, which develops into pneumonia and is then finally capable of becoming a bronchial carcinoma.

Zeolites also can eliminate radioactive material (plutonium). For this reason, the sarcophagus in Chernobyl has been filled with zeolites. Who knows?! There are even plans for using them in Fukushima.

10.3.2. Black Serpent Stones

It is a nightmare to be bitten or stung by a poisonous animal when visiting the tropics. Snakebites kill around 40,000 to 50,000 people each year. 2 % of adults and 20 % of children who are stung by scorpions die from it. The number of people injured by scorpions each year is estimated at 150,000.

The natives in Latin America, Africa, and Asia have been facing this problem for hundreds of years and solved it by healing black stones. These stones are known under various names: Black Stone, Serpent Stone, Pierre Noire, or Pierra Negra. Supposedly the stone is beneficial in treating not only snake bites but also scorpion stings or injuries due to consumption of toxic seafood.

I frequently encountered these stones during my work in an African bush hospital. I learned about them and began to estimate them highly. I never understood their working mechanism, though. My thought was that they probably are capable of exercising a strong suction effect.

In his book "Poisonous Animals", Dr. D. Mebs (79.) mentions the Serpent Stone: "We have a fragment of a polished, partially porous black stone, which must be immediately applied to the bite or sting so that the stone will suck out the poison."

Clearly, Serpent Stones are more than just an aid. Hence, we may ask ourselves how they function. On my search for an answer to this question, I came across the book "Naturopathy in the Tropics", by M. Hirt and B. M'Pia (53.), which describes the Serpent Stone's working mechanism as based on it being a powerful adsorbent capable of rapidly taking in large amounts of fluids. Apparently I wasn't too far off with my speculation about the stone's strong suction. One interesting thing is that the stone sticks to the wound until it is saturated with fluid and falls off by itself, much like for example leeches do. We could say that black stones function similar to leeches. This can be tested by applying the stone to the tip of the tongue: the stone sucks itself to it and when you don't take it off after about five minutes, it can cause bleeding to the tongue.

In contrast to zeolites, which come "ready-made by nature", the Serpent Stone undergoes a complicated manufacturing process. The "White Fathers" from a Christian missionary society managed to uncover the secret of the stones' manufacture. The manufacturing process starts with a cow's thigh bone, which is then processed into a "Black Stone" in various complicated steps. This fact was confirmed to me by a college friend of mine who has worked as a development aid worker in Africa for 22 years.

Anyone who lives and works in Africa for a longer period will encounter these Black Stones. They show that there is still a lot that we can learn from primitive peoples. The complicated manufacturing procedure of the stones shows that people have been capable of preparing highly effective pharmaceuticals for thousands of years. From their way of thinking I took various things that naturopaths have been practicing already for a very long time. Us conventional medical professionals can learn a lot from them.

10.3.3. Microalgae

There is a secret known to the Chinese and many Native American tribes: detoxification with the help of marine algae.

Algae are among the oldest life forms on this earth. Many billions of years ago they formed by photosynthesis and are responsible for 80 % of the earth's oxygen supply.

Chlorella is the only green fresh-water-algae, whose name is due to its high chlorophyll content. The algae contains high amounts of the following substances: vitamins, iron, calcium, potassium, magnesium, phosphorus, and zinc. It also contains a gigantic amount of nucleic acid, proteins (amongst which is methionine, an amino acid), growth factors and a high amount of chlorophyll.

Chlorella has been known to be an optimal source of nutrition and as a remedy for inner purification. After its introduction to in the West it became in high demand as a target of research. D. Klinghardt (64.) was one of the first scientists who occupied himself with this substance. Since he could not realize his goals in Austria (his home country), he moved to the USA where he developed the substance known as Beta-Reu-Rella. He discovered that these algae bind mercury which can then be excreted via the bowel.

Example of therapy after D. Klinghardt (64.):

- Beta-Reu-Rella (in capsule form) 2x to 3x a day, up to 6x a day
- Hepatica and Solidago drops to support liver and kidneys
- Solidago drops: initially, 3x10 or 20 drops before meals, to be increased up to a maximum of 3x40 drops.
- Hepatica drops: initially, 3x10 or 20 drops, to be increased up to a maximum of 3x40 drops.

For CNS diseases, I administer coriander herb in the form of Paracilantro drops, which can pass the blood-brain barrier and remove metals from the brain.

Dosage: Paracilantro drops (coriander tincture)

Initially, 2x5 drops, to be increased up to a maximum of 2x20 drops a day.

Treatment Hepatica and Solidago drops should be continued so that the heavy metals that are released from the brain cannot deposit in other organs and then cross the blood-brain barrier again back into the organ they originally came from.

Today we have Biologo Detox as an alternative to Beta-Reu-Rella and Paracilantro. It was developed by T. Ray and contains Paracilantro and Beta-Reu-Rella in micronized form. In addition, after Uwe Karstädt (61.) it contains:

- Chlorella-Mikro-C, for detoxifying purification of the blood
- Cytoflor TM, to strengthen and rejuvenate the immune system
- Himematsutake, the *agaricus subrufescens* mushroom has been in use in TCM for thousands of years. It has anti-tumorous properties effective for tumors in the abdomen. It also strengthens liver function and the immune system.
- Reishi, strengthens the brain and the nervous system. In TCM it is recognized as a specific mind and heart tonic. Because of its wide area of application and its effective absence of side effects, it has a reputation as the "king of herbal medicine".

I specifically turned to medications without side effect, such as zeolites and marine algae, for detoxification cases where the patient didn't tolerate chemical drugs or during aftercare, when the patient had clearly improved and was ready for a long-term treatment, spanning several years.

Nevertheless, even with medications without side effects we must take care and be sure that we have enough experience with the remedy, in particular when administering to patients with damage through chronic toxic exposure. These patients often react highly sensitive to homeopathic doses or even to only a few drops taken from a naturopathy drawer. In these cases, patients are required to take their medication in the smallest possible doses, under strict supervision by me, within my own praxis.

I would like to encourage our scientists, especially those who work in the pharmaceutical industry, to further develop these substances following the example by Klinghardt, Karstädt, Ray, Hecht, and Straube.

10.3.4. Homeopathy

Most naturopaths also treat their patients with homeopathic medications. I can say the same about homeopathy that I said about acupuncture and naturopathy: it will not perform miracles but, in individual cases, can bring a big and rapid relief to the advantage of both patient and physician. It also protects against the excesses of medical drugs, since homeopathy is much less expensive.

Everybody should experience for themselves how a patient feels better only a few days after taking their little pellets. My first trial patient was the daughter of a good friend of mine, who had problems in school and suffered from skin troubles. In the course of my treatment, her performance in school markedly improved and she passed her exams without effort—achieving a B in mathematics, something that had been unthinkable earlier on. Her skin problems improved but did not go away. Regrettably she discontinued treatment after a few weeks.

Hence, treating chronic poisoning with homeopathic means has its validity today. However, one should keep in mind two things:

1. Due to the vast diversity of noxa, homeopathy has reached its limits. If 100 to 200 years ago a single person was exposed to a single harmful substance and only two or three noxa as little as 30 or 40 years ago, it nowadays usually involves many toxins (although we must keep in mind that these days laboratory tests can find many more harmful substances). Many of these substances and their metabolites are in fact unknown. That which I do not know I cannot treat.

The effect of this plethora of harmful substances is not additive, but involves potentiation of the damage they individually cause to our health. Hence, this spread and potentiation of the effects of these substances must be taken into even more account when following a homeopathic course of action. My teacher, F. Bahr, had this to say about it: "Our development of new homeopathic medications cannot keep up with the rise in new foreign substances." In today's homeopathy, it is often no longer possible to recognize the exact "homeopathic simila" for this diversity of noxa, in particular when one can't commit enough time to the search. In truth, problems of time should not exist in medicine, however, even in homeopathy, this problem is on the increase. The homeopathic physician often needs long sessions with the patient in order to find the right constitutional remedy, taking as much as two to three hours. In order to reduce the time needed for this first anamnesis, I gave patients a question form to be filled out at home. If we don't take enough time, we run the risk of prescribing medication that is not an exact fit to the individual's type of constitution. Through my close cooperation with many homeopaths, I was able to determine that some of them prescribe their patients as many as eight or even more different substances. Hence, they make the same mistake that conventional medicine makes on a daily basis: I never prescribed more than one homeopathic remedy at the same time. This approach has one major benefit: when it helped, I knew that I had the right remedy.

2. Many homeopaths don't know that homeopathic remedies also have side effects, something which C. Fritzsche (32.) points out in his paper on side-effects in homeopathic and allopathic medicine. My advice to all physicians who apply homeopathic methods: before prescribing any drug, be sure to check its ingredients! The rules for vaccinations and medical drugs also apply to homeopathic medication.

10.3.5. Acupuncture

The word "acupuncture" comes from two latin roots: "acus" (needle) and "punctura" (prick).

Acupuncture is only one part of Traditional Chinese Medicine (TCM) and has been applied to animals, humans, and even plants, for more than

4,000 years. According to Chinese philosophy, our life energy, Qi, flows along certain meridians that span the human body. Physical health, in the sense of TCM, means that the body has established an equilibrium, with Yin and Yang balancing each other. The image of a balance is an old-time favorite for illustrating this equilibrium, or the lack of it. Lack of equilibrium means illness. G. Maciocia (75.) in his book "Fundamentals of Chinese Medicine" describes TCM in extensive detail. It is an exciting and educational book that offers deep insights into ancient Chinese thought—insights that are still valid today.

In China and many other Asian countries, acupuncture is used even to assist surgery to some extent. If this is indeed possible, we need no other proof of its validity.

In ancient China, acupuncture was limited to the torso and the limbs: applying acupuncture to the ears was essentially developed by Nogier and Bahr (91., 92.). It was, in fact, Nogier who one day discovered that if he applied acupuncture to a point in the middle of the earlobe—called the eye point—he could improve the eye's vision. When searching for further acupuncture points, Nogier and Bahr found that every single point in the body had a similar point located on the ear. These findings led to the development of ear acupuncture.

I have occupied myself with acupuncture since 1980, at first only through visiting courses and with literature study. For ten years I refused to believe that sticking a needle into someone is capable of healing—for me as a natural scientist this is simply incomprehensible. However, during the past 12 years of my medical practice acupuncture became a method that I use on a daily basis and which I would no longer want to miss from my portfolio of treatments. In combination with Interference Diagnostics, I encountered several cases of spontaneous recovery, as surprising to the patient as to myself, which produced a lot of happiness and thankfulness.

My first case concerned a ten-year-old boy, my caretaker's grandson, who suffered from a stuttering disorder. After as little as two sessions his situation improved. After his third and last visit, he was almost capable of speaking normally again—something miraculous had happened. I frequently asked my caretaker how her grandson was doing. Both she and the boy's parents were satisfied with his speech. The boy positively flourished. In my capacity as Spokesperson for the "Nutrition and Health" quality committee, I held lectures about acupuncture, which evidently infected more than one of my colleagues.

10.4. Environmental medicine is holistic medicine

However, environmental medicine is more than this—it always is holistic medicine, as described strikingly and comprehensively by J. Beuth (6.) in his book "Healing Cancer Holistically". Beuth has been leading the Institute for Scientific Evaluation of Naturopathic Methods, at the University of Cologne, since 1999. On first reading his book, my spontaneous thought was that every university clinic needs to associate with such an institute.

That which is of importance to cancer applies to all environmental diseases. Beuth attributes a high value to balanced, individually tuned nutrition, to sports, psychological counselling, vitamins, and trace elements. He also considers strengthening the immune system very important. To achieve this, patients need to be tested for different sensitivities, as I described in chapter 9.8.4. In the end, it is the body's own defenses, which in turn depend on bioenergetics, that decide about health and disease. This current insight is, however, largely unknown. Hence, it is even more surprising that this insight into life energy is quite common under many naturopaths of the younger generation. This gives hope for the future and tickles the scientist's curiosity.

Two people have been exposed to the same noxa. Question: why does one of them become ill whereas the other doesn't? A strong immune system protects against disease! I would like to express this in saying: "Take your physical health into your own hands." In particular, this involves a healthy lifestyle, capable of strengthening the immune system. This is something that cannot be done by the government, nor is it the government's task. Nevertheless, the government should create the preconditions.

11. A healthy lifestyle: what can we do ourselves?

11.1. Introduction: what does "healthy" mean?

In the previous chapter I explained that when recording one's own medical history, one should also examine one's way of life. If an unhealthy lifestyle turns out to be the root of all evil, even the best and latest types of environmental-medical therapy will not help. Rather, one should change their habits, their way of life. But where to start?

> *I here repeat my wish for everybody to take their own health into their own hands! Our health is our most precious possession and we should take care of it and nurse it—each of us during our own journey through life.*

In the end, the journey is the reward. On this road, as Anselm Grün (43.) proposes in his book "What Should I Do?", there are no rules, nor is there any advice to be found. Grün writes: "Advice is often vice that is added to our trouble."[36]

Elsewhere in the book he continues by quoting a Finnish proverb about advice: "Good advice is like snow. The more silent it falls, the longer it remains." Both Grün and myself like this line very much. Hence, all my remarks are only cautious recommendations that are based on my own experience, as reflected in this book's title.

About ten years ago, a senior citizens club asked me to hold a lecture on the topic of "growing old in good health". This really is not an easy task, even for a doctor, and hence I made an in-depth study of the available literature. During my research, I encountered the following quote:

> *"If the spirit is young, age doesn't matter."*
> Elliott Carter

Once again, mind controls matter. I continued by researching the concept of "health". What does it mean to be in good health? To my surprise,

---

36 Translator's note: the original text plays on the German word for "advice". As always with wordplay, a translation can only attempt to convey a whimsical original.

I didn't find an unambiguous explanation. One may summarize it by saying that someone is healthy if they are not suffering from any physical or mental symptoms. When I continued searching, I encountered the following definition:

> *Healthy is that person who lives in harmony with themselves and their environment.*

This definition is the one closest to my own way of thinking, for it includes the fact that everybody can decide for themselves whether to feel in good health or not. It also includes the environment—something which, in my understanding, is very important.

Hence, we are once more back with environmental medicine and its first and decisive question: do I have an environmental disease? Am I living in harmony with my environment? If I must answer this question negatively, the next question must be how to become fit—and how to remain fit—both physically and mentally. Here, nutrition plays an important role.

11.2. Nutrition

At one point in my life I was suffering toxic effects from exposure to palladium. After dental restoration work and ensuing detoxification, my health improved markedly. Since we are, however, continuously exposed to harmful substances and hence suffer from the corresponding exposure, I have underwent a detoxification regimen once or twice a year ever since, alternating between e.g., Biologo Detox and Toxaprevent (zeolite).

An additional measure I took was to change my nutritional pattern. I am not a vegetarian, but in weekly steps reduced the amount of meat in my diet to 20–30 % and have been eating more vegetables (fruit, vegetables, potatoes, rice, spelt and rye bread) and drink a lot, mostly pure tap water. This last I can however only recommend if the local tap water tastes well and is of good quality. In the big cities, this precondition is often not satisfied. The problem of "nutrition and clean drinking water as cornerstones of our health" already starts here.

Especially when it comes to nutrition, everybody must find their own way. One should also take care to consume sufficient quantities of vitamins, trace elements, and minerals. Since German women appear to drink insufficient quantities of fluids, I recommend my female patients to drink one more glass (200 ml) more than their thirst merits, in particular on warm summer days.

Chocolate, sweets, and all sorts of treats that are on offer are and remain indulgences for me, which I eat, and allow myself to enjoy, on special occasions.

11.3. Sports, exercise, gymnastics

It is a well-known fact that exercise and sports help prevent cardiac infarctions and cancer. They are part of a healthy lifestyle and should also be part of any holistic course of treatment. Both stimulate blood circulation and perfusion, as well as our metabolism, digestive system, and the elimination of waste products and foreign substances—including toxic substances. I am a person who likes to move a lot. I perform gymnastic exercises for about eight minutes every day. Three to four times a week I go jogging for 30–40 minutes, or longer in summer. Fresh air and nature simply are tonics for my body, mind, and soul. When I'm exercising outside, I feel in harmony with the environment and with myself, and I prefer those walks above any TV show.

The Way of St. James
I had a similar experience when I walked the Way of St. James. I started out in Le Puy-en-Velay in the Massif Central in French, in 2009, when I was 71 years old. It was my intention to make a long-time dream come true. Since my time was limited, due to family duties and private reasons, I walked only 10–14 days each year. By now, I managed 45 days (in four stages), spanning 1,400 km of walking, and arrived in Santiago de Compostela on April 6, 2013.

Whenever I walk the Way, I go by myself, just to make sure I have a little more time for myself. Nobody disturbs me and my solitary hike is good for me. I can pause wherever I want, can lie down in the grass and look at the sky, and dream and meditate. Whenever I go out wandering, it takes me only a few minutes to forget everyday life. I am able to enjoy nature and can put up with—even enjoy—the company of friendly fellow pilgrims. Each of them has their own story to tell and is enjoying the freedom and full-range of the countryside. From this, we gain new strength. In short, when walking such a pilgrim's road, one is closely connected with nature. What more do we want?

I am always surprised how few necessities we need on our road. Including my backpack, I carry as little as 6.4 kg. When I walked the first stage, this still was as much as 12 to 13 kg. Once again, this increases my feelings of freedom.

11.4. Massage

What more can I do for myself that does not involve taking medication? I allow myself to get a massage every now and then. Specifically, I am speaking of Western-style massages, as developed in our industrialized countries. Amongst those, we count classic massage, foot reflexology, and the Breuß-massage.

It would spring the bounds of this book to detail each individual type of massage and its workings. Nevertheless, I would like to briefly discuss their manifold mechanisms of action.

The aim of massages is to mechanically influence skin, connective tissues, and musculature via stretching, pulling, and pressure forces. Its effects spread from the treated location through the whole organism, including the psyche.

The classical massage technique dissolves muscle tension, purifies the tissue, and stimulates blood perfusion and metabolism. Foot reflexology works as follows: all of our inner organs as well as all of our locomotor system are reflected in so-called reflex points in our feet. When these points are massaged, this has a beneficial effect on the whole organism and helps harmonize energy flows through our body. This technique can be used for preventative as well as therapeutical measures. Breuß-massage involves a gentle, energetic back massage, which strengthens the spinal column and frees dammed energy flows. It also rejuvenates any neglected intervertebral discs.

Besides the Western massage techniques, there also exist massages that were developed in the Middle East and in Asia, many thousands of years ago. Examples come from India, Tibet, China, Thailand, and Hawaii, and include for example Ayurveda massage, Hawaiian massage, and Thai massage. Today, we call these techniques "health-spa massages", but they are much more than that: they too help stimulate blood perfusion, loosen tensions, and relax body, mind, and soul. Since it increases lymphatic circulation, it also helps wash out waste products from our tissue.

During a massage, I feel like in the front yard of the garden of Eden. Afterwards, my head and thoughts are crystal-clear—an effect that lingers for days and weeks. Hence, massages are another way of bringing people back in harmony with the environment. If you just want to forget the everyday world for a little time, then allow yourself a massage rather than taking pills.

11.5. Psyche—the significance of our souls

I tend to avoid using the word psyche and prefer speaking of the soul. Science increasingly recognizes the extent to which our thinking influences our action, feelings, and hence also our health. Psychological stresses can cause genetic damage and weaken the immune system. Hence, body, mind, and soul are much more closely connected than was assumed before.

The nervous system/psyche and our immune system mutually stimulate, or inhibit, each other by the use of various chemical signalling substances such as serotonin, our happiness hormone. Happiness and laughing indicate a positive mood, which can now be measured in laboratory tests, and they activate the immune system. Grief, tiredness, disappointment, and exhaustion have the opposite effect and open the door to depression. When taking my state exam in psychiatrics, I was asked to define the term "depression". My answer was: "Depression, to me, is continuous, abysmal sadness."

For me, the test was then over, since I clearly had given the correct answer. I always have to think of this when patients with environmental diseases tell me their many-year-long ordeal, which is characterized by phases of depression and feelings of loneliness.

Why am I mentioning this? In the past, depression and loneliness were problems that in particular elderly and diseased people had to deal with. For about the last 20 years, this has been quite different. Today, younger people are increasingly suffering from these problems—most of them have environmental damage. This damage caused them to become unemployed at an early stage, often even making them fully invalid for work. They lost their jobs, including fond colleagues and friends. They end up feeling redundant and abandoned by the world. Their loved ones don't know what to do and also retreat from them. Their family physician frequently also gives up and sends them from one specialist to the next.

This starts a medical odyssey without any end in sight, which usually ends with being categorized under "psycho-diseases" and being pumped full with psychopharmaca, without the actual cause being clarified. As a consequence, these patients go downhill even further. With this, I arrived at those who are in most urgent need of help. They are the ones that I particularly target with my book. They too can be helped.

According to A. Grün (42., 43., 44., 45.)[37], when one finds oneself in such a situation, the following factors are important to one's health:

- Reconciliation with the past. Forgiving everyone who ever harmed you.
- Accept your own limits. This means to accept who and what you are, including your strengths and weaknesses. Everybody has strengths, which should be mobilised.
- Learning to cope with loneliness. Use this forced time of tranquility to listen to your own body and tune all your senses towards yourself. Regard your disease and your loneliness as a challenge towards changing your life, to bring it back in harmony with the environment and make peace with yourself. That this is possible, I experienced in myself and in many of my patients.

Anselm Grün is a padre in the order of St. Benedict. He studied theology, philosophy, and business administration. He is the most widely-read contemporary christian author. I have listed several of his works, which I studied when preparing for my lecture, in the bibliography. They discuss healthy living and meditation and are worth reading because they are easy to digest but at the same time convey telling information.

For more than 40 years I have been making visits to patients, both privately and in my capacity as a doctor, even though today I am retired. These visits are very important to me and to the patient, they are a pleasant and usually joyful change of scenery. I always tell them about Anselm Grün or Hermann Hesse. To the second of those, accepting your own loneliness is the road to enlightenment.

---

[37] It is not my intention to plug A. Grün. However, especially in critical situations, it is worthwhile to seek help in books such as his.

The German philosopher Arthur Schopenhauer takes it one step further when he writes: our youth should work at learning to bear loneliness, because it is a source of happiness and tranquility.

Coping with loneliness—finding peace in a monastery
For many years, once or twice a year I have been spending a week on sabbatical in a Franciscan nunnery. The peace and quiet that this place radiates time and again soothes me. As soon as I first walk through their wide corridors with their many beautiful antique furniture, a feeling of quiet comes over me. Freed from everyday tasks, worries, and responsibilities, all my senses turn towards reception. My mind starts to sparkle and without these sabbaticals, this book would not have been possible.

11.6. Meditation

Meditation has been a part of my daily life for years. It is as essential to me as eating, drinking, and sleeping. I would not want to miss these 15 to 20 minutes of inner peace and quiet—they carry me to a place of unlimited possibilities. According to J. van Praagh (119.), meditation is the simplest and most reliable way of looking inside yourself. It leads to a restoration of the equilibrium between the emotional, mental, physical, and spiritual self. Here, one can find safety and security, something very important in today's world of permanent wars, catastrophes, and crises, which seems to be increasingly bursting at the seams.

In the preface to his book "meditation for inner strength and joy of life", van Praagh writes: "During meditation, one activates cosmic energies. These energies revitalize and enlighten the spiritual centers in our body. This way, the light of unconditional love deep inside yourself is ignited, which radiates ever more brightly with every time one meditates. The deeper the changes inside oneself that this light causes, the stronger one can positively influence other people."

For me, meditation has become so important that it is the first thing I do every day. During these minutes, the structure of the coming day becomes clear, it prepares me for the day ahead. Looked at it in this way, it provides harmony with the environment at a higher spiritual level. Loneliness and meditation are wonderful complements. Metaphorically speaking, loneliness to me is the wide river that spills into the endless sea that is meditation.

Why am I mentioning loneliness? What does this have to do with environmental medicine? Today, the answer is, regrettably, a lot. Many patients with environmental damage feel abandoned and misunderstood, even by their own loved ones. They are placed in the "psychiatric" corner. To this we can add financial worries and these cases often involve young people, some of which at age 30 or 40 already have to fight for their pension.

11.7. Hobbies

I wish for every human being to have at least one hobby that can be done outside and one inside—an activity that you love doing and which balances out that which you do professionally. You should already find such a hobby-horse in your younger years, so that after retirement something remains that is capable of giving you pleasure. But hobbies can also be started at a more advanced age. Essentially, age does not play any role.

When I was 13 years old and had to stay in hospital for eight weeks with a broken leg, I filled practically every waking hour with playing chess. It turned out to be a good thing that it already had been a hobby of mine from since I was a little boy. The nurses moved my bed from one room to the next and from one ward to another. Time flew, and I ended up grandmaster of the hospital. As a side-effect, I was able to teach many patients, including children, to play chess. The youngest of them was seven years old and he quickly became the vice champion.

Which hobby should one consider? This is a decision everybody has to make for themselves. As late as at age 60, I fulfilled a childhood dream and started taking piano lessons. To this, I added organ lessons when I was 65 years old. Playing music trains the brain and the memory, and supposedly even helps to lives longer. I do not know if this is true, but to me it is in any case enormously beneficial.

Also at age 65, in order to please my second-eldest brother, I joined his skiing club and learned how to do alpine skiing. Ever since then, I join him and our club for a week-long skiing trip to the Alps. To sit in front of an Alpine hut, under sunshine and minus 15 degrees Celsius, eating hot soup at 1,000 meters above sea level … that is balsam for body and soul. Up there, one feels even more closely connected with nature.

Among my other hobbies, I count hiking, biking, reading, and gardening. Besides those, development aid and science have been my trusty companions ever since 1964. You might say that this is quite a lot of activity for a 74-year-old. I am not listing all this to impress you, but I would be delighted if I could incite even only one reader to pick up a hobby—the more the merrier.

11.8. Bio-meditation—bioenergy: a higher-level therapy

We can still learn a lot from the famous Greek physician Hippocrates, who left us the saying:

"The physician helps, but nature heals."
Hippocrates

What Hippocrates is saying is that no human being can heal another one—only the body can heal itself. A physician must provide the patient with the necessary information. However, if not even a doctor can help, then we must assume that our medical machines are even less capable of helping, no matter how modern they are.

I stressed several times that we have arrived at a crossroads as regards our way of life and way of thinking, characterized as it is by the industrialization of our world. This also applies in particular to our modern natural-scientific medical science. Traditional therapeutic methods speak about life energy. According to V. Philippi (97.), this term means energy that is created by biochemical processes that take place within our bodies (conversion of nutrients into energy).

In his pamphlet "Bioenergetic Meditation", Philippi introduces a "higher" form of energy, which finds its basis in bio-meditation (98.). He writes: "Bio-meditation uses a form of energy that works much more intensively and at the same time flows through everything around us. It is the force that created the basic elements of water, fire, air, and earth, as well as humanity itself and everything that lives: it keeps all of us alive. This energy now is the fifth element, the power and love of God. I call this "bioenergy".

This way of thinking completely captures the spirit of Traditional Chinese Medicine, an approach that I have learned to value for its farsightedness. One can sense this way of thinking also when reading B. Lipton's book

"The Wisdom of Your Cells—How Your Beliefs Control Your Biology", which I already mentioned before.

So far, these higher-level approaches are not taken into account within our strictly scientific way of thinking and acting—they cannot be captured even with the highest level of randomized double-blind studies. Philippi also discusses the three pillars of life: thankfulness, forgiveness, and acceptance. The three of them have been part of my daily life for many years and provide me with necessary peace, serenity, and endurance. Here, we also come full circle.

Life energy has dominated the 20th century. It would be desirable if bioenergy would emerge from the shadows for the good of all of humanity.

I am very much aware that many readers will greet my description of ancient therapeutical methods on the one hand and my impetus towards new way of thinking on the other hand with a shake of the head. To them, I would like to repeat what I already mentioned in the preface: in this book, I present my own experiences, which are to a large extent corroborated by scientific studies.

12. Course of action: the seven steps

We humans need targeted instructions whenever we want to introduce new therapeutic steps. In the rest of this chapter, I would like to provide you with such instructions.

Step 1: detailed anamnesis
Our aim is always to find what caused a disease, and it is the most important decision that has to be made. Only if I decide right, I can diagnose my patient with certainty. For this, we need a comprehensive record of the patient's medical history, which consists of the following aspects: known diseases in the patient's family (parents, siblings, other relatives and ancestors), the patient history prior to the disease, professional career, how a patient spends their free time, nutrition, hypersensitivities, allergies, prior surgery, etc. With the help of this record I am able to find 90 % of what caused the disease. This important preliminary work, which sets the course towards a correct diagnosis, can be done by the patients themselves. Hence, they are largely capable of providing an answer to the question: do I have an environmental disease?

With this knowledge about themselves in hand, patients can better answer the questions asked by their attending physician and ask questions in return. It also enables them to better monitor any follow-up examinations for the extent to which they are sensible and correct.

Step 2: conventional-medical examination
The next step in investigating a patient's disease can only be taken within the bounds of conventional medicine. As I already extensively argued before, it is not possible to achieve anything without carrying out these examinations. For this, a detailed medical history, prepared by the patient themselves, is highly helpful to a family physician. Using an environmental-medical questionnaire makes everybody's work easier, and has the added advantage of reducing the risk of oversight.

Step 3: making an appointment with the family physician for a special consultation.

I always recommend my patients to make an appointment with their family physician for a special consultation, so that he or she can make enough time for the patient and for reading their medical history. During such a consult, doctor and patient get to know each other better, which creates trust and raises hopes. Both of these are essential in order to get well.

Today we know that psychiatric diseases can cause genetic damage. On the other hand, we also know that trust and fresh hope can move mountains, figuratively speaking. Why would those factors not be capable to also heal genetic disorders? I strongly believe they can, even though I cannot prove it. In any case, many of my patients told me after such a special consultation that this by itself already increased their courage and gave them hope. After treatment they were doing markedly better and their life once again seemed worth living. What more does one want? No technology can boast such an achievement.

Step 4: referring the patient to an environmental physician

If the cause of the disease cannot be found by conventional means, I recommend referring the patient to an environmental physician. You can find some addresses at the end of this book.

Step 5: material analysis, carried out by a dentist or a dental technician

If the patient wears any dental restoration material in their mouth, something which is almost always the case, one should obtain a material analysis from one's family dentist. If dental restoration work is indicated, the attending physicians should follow the ten rules that I outlined earlier: first remove the material and then detoxify the patient.

Step 6: inform your health insurance about any environmental-medical examinations

I recommend my patients to always inform their health insurance company before commencing any planned examinations. First of all, this enables the insurance company to consider any costs connected to the treatment well in advance. Second, it is useful to inform them of the existence of environmental diseases in the first place. If one neglects to inform them, the same will happen that I experienced myself during a visit to the KV Dortmund, where both its chairperson and the representative of the insurance companies avowed that they had never seen a patient with an environmental disease.

When preparing a written justification, it must be expressed clearly that all of the conventional medical examinations and treatments that have been conducted so far had any success. Every family physician, internist, or hospital doctor can issue such written confirmation. It is best to present the insurance company with the findings from all examinations at the same time.

Step 7: legal proceedings
If it comes to legal proceedings, one should consult a lawyer with relevant experience in the field of environmental-medical law (see the appendix for an address).

In the end, it is the environmental physician who must decide what further examinations and treatments should be carried out for an individual patient. This too should be communicated to the health insurance company in the written statement.

At this time, the biggest and only chance for sustainably maintaining our health—and that of our progeny—that I can see is provided by holistic medicine without side-effects, which includes both conventional medicine and tried-and-tested traditional methods of healing.

If you want to master the present, then you must read the past.
If you want to plan for the future, then you must analyze the present.

13. Critical assessment and conclusions

13.1. Introduction

All the great cultures in human history always regarded the environment, animals, plants, and humans as a whole in which all parts mutually influence each others. This fact seems to be forgotten by us people from industrialized nations. We exploit nature and destroy our valuable plant life, animal life, and also each other as human beings. One visible sign of this is the steadily increasing number of chronic diseases and cancer.

What are now the medical, legal, moral, and ethical conclusions that we must draw, without losing sight of the vast area that is our social politics?

13.2. Medical assessment

Harmful substances from the environment and their relevance to our health
The aim of presenting case histories is so that anyone affected may recognize themselves in them. They are already sufficient to show that harmful substances from the environment are jointly responsible for numerous chronic diseases. Harmful substances can cross the blood brain barrier and cause neurodegenerative as well as psychiatric damage. They can also cause benign and malignant tumors to develop. The studies that I conducted have provided us with insights above and beyond the aforementioned.

There is, however, clearly no concrete connection between exposure to a specific metal and commonly made pathological-histologic diagnoses. The latter don't tell anything about exposure to metals, and hence about the possible cause of a tumor. Hence, today it does not provide the key to successful therapy.

The same probably applies to all other noxa. This fact is remarkable and will probably surprise—and alienate—many conventional physicians, pathologists in particular. Nevertheless, the diagnoses obtained with the above in mind are not superfluous, since they can perhaps help see a connection between metals and any pathological-histologic findings. Maybe as early as in the near future, we can use electron microscopy to not only look into the atoms of inorganic substances, but also show presence of metals inside the cells, the mitochondria, cell nuclei, etc. This would possibly also provide visible evidence for their role in the development of cancer and other chronic diseases.

I would advise any colleague who is critical, or completely disapproving of harmful substances as causes of chronic diseases and cancer, to give me proof to the contrary. Until then, I recommend to examine any patient who suffers from a malignant tumor for exposure to noxa. This must be done in a goal-oriented way, in accordance with the patient's medical history. This recommendation also applies to anyone who suffers from a chronic disease.

If tissue contains nickel, it should not be subjected to radiation treatment. It is common for irradiated tissue to develop an angiosarcoma after the treatment. It is not yet known what causes this and, according to my knowledge, also has never been studied. Nickel allergy is the most frequent allergy. If nickel-containing tissue is irradiated, this causes a strong provocation that may lead to renewed cell proliferation, which can therefore go on to cause cancer. I observed several such cases. What is valid for nickel probably also applies to all other metals. However, apparently nobody has noticed so far, nor has the question even been studied—it would destabilize the whole method of radiation treatment before or after cancer surgery. Additionally, radiation therapy is not a causal therapy.

My studies now showed that not only the tumorous tissue, but also the surrounding tumor-free tissue is exposed to carcinogenic and potentially carcinogenic metals. From this it follows that one should fundamentally never subject the affected organ to radiation treatment. This coincides with what I said in chapter 6.5. about the connection between metals and electromagnetic waves.

Acute metal allergies can lead to impairment of postoperative wound healing. I would like to recommend my surgeon colleagues to refrain from using tackers in the future. Tackers are used to close the abdomen after surgery or to reconnect both ends of the bowel after part of it is removed (bowel resection). These clips are made from precious metal alloy and contain carcinogenic and allergenic metals such as nickel, chrome, iron, carbon, and other metals depending on the manufacturer. Nickel is the most common allergen.

Metal poisoning should also be considered in any pediatric disease, such as ADD and ADHD, neurodermitis, eczema, ichthyosis, migraines, and all types of bowel diseases. This advice is backed up by the 80 % recovery rate after detoxification treatment.

In general, harmful substances from the environment clearly play a large role in the development of diseases: hence, if professional therapy is conducted, the chance of successful recovery is, in my experience, 80 % to 90 %.

In this context, I would like to mention that in the amalgam verdict from May 31, 1996, the same 80 % recovery rate was found for the total of all the patients involved. This corresponds exactly with my own results.

The relevance of dental restoration material

The study on the danger to health posed by dental restoration material shows that most of the people who are suffering from an environmental disease do not wear just one single material, but often have a smorgasbord of different heavy metal and precious metal alloys. Many of them also had budget gold crowns in addition to the gold ones. More than 90 % of patients with malignant tumors wear two and more alloys and 74 % of them has more than six amalgam fillings.

Even proponents of amalgam admit that accumulating metals are dangerous to our health. This suspicion is strengthened even further by the results obtained from dental tissue, which showed exposure to a multitude of metals.

All participants in my studies wore dental restoration material.
Bottom line: a full generation is at risk of creeping metal poisoning.

It is not just amalgam, but also other dental restoration materials, as well as synthetic materials, that carry a high risk to our health. Hence my call for better testing procedures to check these materials for biocompatibility.

Intrauterine damage to the fetus

Our results further show that probably not only metals, but also other substances can be transferred from the mother to the fetus. These substances then may be causally responsible for many chronic diseases in the child that carry through into adulthood. Hence, the embryo's exposure to harmful substances often already begins inside the mother. Babies are born already poisoned. To know this and to have to write this hurts!

Harmful chemicals and metals can be transmitted from the mother to the fetus and cause severe health damage—even miscarriages and stillbirths.

This fact has been established more than 20 years ago. However, it still plays practically no role in pediatrics and obstetrics. There is a large need for clarification and research into this field!

My recommendation is to examine babies and children that suffer from diseases and impediments of unknown etiology for metals already at an early stage and, if necessary, start appropriate treatment. At the same time, the mother and even the father should also be tested for exposure to any harmful substances.

I described in chapter 8.6.3. how mothers who suffered a miscarriage or stillbirth were exposed to harmful substances that had been transferred to the fetus or the embryo within the uterus. Hence, we must assume that miscarriages and stillbirths are caused by noxa. There is yet another reason to make this assumption: after one stillbirth, I measured strongly elevated levels of mercury inside the placenta—both parents had been exposed to mercury. After detoxification treatment, they got two healthy children.

Which final conclusions can we now draw? I recommend to conduct a postmortem examination in all cases of miscarriage and stillbirth and to test the tissue for harmful substances, in order to find the cause of the tragic event. Here, I would once again like to point out the findings made by the Munich University pathological institute, which also showed exposure to metals in the brain of stillborn babies and miscarriages. These findings provide proof for toxic damage.

In such cases, we must re-ask the question as to when any possible genetic damage was done. There are three possibilities:

1. Transmission of the genetic defects from mother to fetus.
2. Intrauterine genetic defects (caused by, e.g., harmful substances).
3. Genetic defects that develop during the course of life.

Due to the high medical significance of intrauterine fetal damage, I would like to take this opportunity to ask pediatric doctors, gynaecologists, and obstetricians to confront this issue, both as a doctor and as a scientist, to a larger extent than they do now. The childrens' parents will thank you for it.

Health dangers posed by organic mercury and tin compounds

The most frequently type of exposure that is found in urinalysis is not to mercury, but to tin (inorganic as well as organic). This is remarkable, because tin is many times more toxic than inorganic mercury. With this, there exists a gap in our scientific research, since in the past health risks due to organic metal compounds have hardly ever been studied, despite their up to 1,000-times-higher toxicity in comparison to inorganic metals. Organic metal compound doses in the nanogram range are already highly toxic. These compounds should be subjected to increased attention through regular testing. This is particularly true for the organic compound ethylmercury, which is added to vaccines as booster or as a disinfectant (adjuvant) and which can cause vaccine damage—fact that is hardly known.

Comparisons between the 360 patients with and the 377 patients without environmental health damage who attend my general medical practice show an up to eight times higher frequency of chronic diseases in the group with environmental damage as compared to the control group. This is a highly significant result.

The participants in both groups of patients with rare diseases and benign tumors showed exposure to metals, as measured from empty-stomach urinalysis. In the study on cancer patients, I found two metals in their urine and in the two tissue samples I found five and six toxic metals, respectively. From those metals, at least one of those found in the urine and two that were found in the tissue samples are also carcinogenic.

Metals and harmful chemicals can pass the blood-brain barrier and cause brain tumors, as can be seen in the two case studies that I presented of the child with the germ cell tumor and the woman with the rapidly proliferating glioblastoma.

Immunological and genetic damage certainly are also responsible for developing cancer, but they are not its true cause. The first triggers for cancer are metals and other noxa, such as chemicals, fungi toxins, as well as electromagnetic and radioactive radiation. These factors cause genetic damage and destroy the immune and endocrine systems. Today, both animals and humans, particularly those who live in metropolitan areas, are fighting with this toxic cocktail. There is an economic boom also for animal hospitals.

All the examples and studies that I presented show a causal connection between harmful substances and chronic diseases as well as cancer. Here, metals are among the most frequent triggers. It cannot be a coincidence that carcinogenic and potentially carcinogenic metals are most frequently found in the urine and in both tumorous and surrounding non-tumorous tissue samples taken from precisely those cancer patients. Here, the chain of indicators is closed and the vicious circle that I described in chapter 8.4.4. takes effect.

As long as new scientific results do not provide any proof to the contrary, we must hold on to this thesis. To not acknowledge this opinion only because there are no reference values that apply to organs is unmaintainable in light of the unambiguous findings. In the end, final proof is extremely rare in medicine in the first place. For me, the experience I gained over this period of 25 years is rather more important. As pointed out by B. Lipton (73.) you can control your own genes. Hence, we should draw the necessary conclusion as soon as possible. In my opinion, to wait and see for longer, without presenting alternative solutions or providing proof to the contrary, is no longer justifiable.

According to a DPA press release (21., 22.) issued on December 7, 2012, about 220,000 people die of cancer in Germany each year. This amounts to 26 % of the yearly death rate. According to M. Rath (99.), the worldwide death rate due to cancer is 7.5 million. The press release did not mention the number of people who develop cancer. Why not? Well, it might be as high as twice the number of deaths. That which applies to cancer is also valid for other diseases of affluence. Cardiovascular diseases alone amount to 348,000 deaths each year—40 % of the total number of deaths.

Even proponents of conventional medicine must admit that despite their efforts and despite the many hundreds of billions of euros that have been invested over the past 20 to 30 years, the incidence of chronic diseases and cancer has increased dramatically. Something must be wrong here!

In light of this fact, it is about time that university clinics start conducting large-scale epidemiological studies.

If they neglect to do this, they lose their right to question this thesis. Of course, introducing reference values is a precondition for such studies—norms are a general rule in almost all fields of medicine and they are necessary to provide a basis for discussion.

A demand for reference values for metals in human tissue

The postulate about reference values becomes even more justified when we take into account the fact that both tumor-free and tumorous tissue from cancer patients seems to show markedly higher levels of exposure than the control group of healthy patients from Ionescu's study (57.). With a factor of 13, the level of exposure in comparison to this healthy group also is rather high. Clearly, in cancer cases all of the affected organ is exposed to metals. The high exposure levels in the surrounding so-called tumor-free tissue might explain recurrences in the same organ. From this it follows:

If a complete organ is exposed to metals, completely removing the tumor is not sufficient for preventing a recurrence.

The continuing exposure to metal might even be the explanation of metastasis. After an operation, depending on the scale of the surgery, all or part of an organ that is exposed to metals will no longer be available as a body store. Hence, the body seeks a new store within the same or another organ. There exists an interesting parallel with our everyday life in how we treat our toxic waste disposal sites.

In cancer cases, it is necessary in principle to also use DMPS provocation testing to test for presence of heavy metals in the patient's urine. If the test result is positive, tissue samples should be taken from the affected organ and, possibly, also from other, healthy organs. This is technically possible, since only a small amount of sample (puncture) is already sufficient. It would be best if any causal connections could be further established in advance, in order to convince even the last few remaining doubters.

Because of all of the above, introducing reference values for metals in human tissue is imperative.

In all of this, I am aware that in individual cases of exposure to multiple different metals these reference values have only limited significance, due to the fact that different toxins mutually reinforce each other. In his 1990 book "The Boundlessness of Reference Values", A. Kortenkamp pointed out that determining and regulating reference values does not take into account the possible interactions between many substances. (93.)

Where else can we observe the special role played by metals?
All of my Lyme's disease patients as well as all who suffered from allergies or were hypersensitive to mobile communications and other types of radiation showed exposure to metals. These added risk factors had caused the cup to overflow. This, one also should know. Detoxification treatment managed to achieve a marked improvement, even complete recovery. Apparently, the body was again capable of dealing with other noxa.

Obviously these other factors were also deactivated, to the extent that this is possible. However, this will not always succeed.

New diagnostic and therapeutic possibilities.
For all chronic diseases, diagnostics and therapeutic treatment should be commenced as early as possible. With the help of professional and timely detoxification, we can take the sting out of the development of cancer and chronic diseases, so to speak. This sting propels the processes towards unchecked proliferation—this is called an inflammatory provocation.

As an example, let me recount the fact that in order to speed up its action, metals are added to vaccines, which provoke such an inflammatory reaction. In this case, it is a one-time administration. In the case of a creeping metal poisoning we are dealing with a chronic inflammatory provocation.

How do I proceed in individual cases? For all cancer patients who wear dental restoration material, material analysis should be carried out before commencing the actual treatment. This of course also applies to other chronic diseases for which conventional medicine can no longer provide any help. At the same time I recommend to test for exposure to metals as well as determine genetic and immunologic factors. Further therapeutic measures can then be determined on the basis of these findings. To wait until the pathological-physiological development of malignant tumors is fully decoded might in individual cases have fatal consequences for the patient concerned.

It will still be necessary to radically remove the tumor within the foreseeable future. However, any further therapeutic steps should also take into account the environmental-medical findings.

This demand for environmental-medical examinations and treatments is even more justified due to the fact that radiation treatment and chemotherapy are not causal therapies and often go hand in hand with the most severe side-effects and may even have fatal consequences. Both these therapeutical options wreak havoc on the immune system. They destroy healthy cells and can even cause malignant tumors.

In 2004, the German epidemiologist Prof. D. Hölzel pointed out that chemotherapy has not been able to increase the chances of survival for patients with malignant breast, bowel, prostate, and lung tumors for the last 25 years. J. Blech (11.) reported about this in the Spiegel magazine.

However, there are even more dangers connected with chemotherapy and radiation treatment, which J. Mutter describes (89.). Both treatments cause increased exposure to, amongst others, free radicals. This way, they damage the mitochondria, which, as we know, have already been damaged by the cancer. Like all medical drugs, chemotherapeutic drugs and their metabolites are excreted by the human body and hence end up in our environment and our drinking water where they form, to say it with J. Mutter, a cocktail of chemotherapeutic and other drugs. However, our drinking water is not regularly tested for these substances. Today this is possible and absolutely necessary.

As mentioned before, we today have to deal with a mix of chemicals and toxic drugs as well as their toxic metabolites, which are foreign substances to our human body. Twenty years ago, nobody would have dared saying this out loud. Hence, U. Karstädt is, regrettably, correct when he makes the demand in the title of his book: "Detoxification Instead of Poisoning." (61.)

Here, I do not want to condemn everything that has come before. In fact, we owe the pioneers of chemotherapy a lot of respect. They have fully committed themselves to these therapeutic methods at a very difficult time for tumor research, as described by for example S. Mukherjee (84.) unrivalled book "Cancer: King of Diseases". Chemotherapy has gained considerable success in particular in treating children with leukemia and for various tumors of the gonads and bones.

Now, because numerous new scientific studies show that harmful substances are crucially responsible for the development of cancer and many chronic diseases, the time has come for conventional medicine to absorb and consider these new insights.

In chapter 10.3., I consciously discussed medications without side effects in somewhat more detail. With this, I want to show that our ancestors have been capable of developing medications according to their needs for hundreds and even thousands of years. Regrettably, with increasing industrialization these insights have been lost more and more and have never been developed any further.

The development of ever newer chemicals and medical drugs, none of which are sufficiently tested—or not tested at all—for tolerability, has created diseases and genetic damage that can no longer be treated with medications without side effects.

By now, 70 % of the population of Europe suffers from genetic damage. Which consequences should we draw from this? We should continue where our ancestors stopped and, rather than develop toxic chemicals, further develop the available natural treatments so that they can truly become drugs without side effects. This could only be beneficial to our contemporary medical science, to the good of everyone.

If we proceed along this road, our healthcare system will once again be affordable and there will no longer be the current two- or three-tier-system. The scales will be balanced and social peace will be kept. The call for safety and more transparency in medicine will be heeded.

In contrast to S. Mukherjee, whose book is suffused with resignation, I want to use this book to invite everybody to look at the future positively, and I consciously appeal to the power of positive thinking.

When we do all this, we will be able to answer the question as to the cause of chronic diseases and malignant tumors, and step by step defeat them like we defeated many of the infectious diseases that have plagued mankind for many centuries and even decimated our numbers—I am thinking of the plague, tetanus, poliomyelitis, and yellow fever. Radiation treatment and chemotherapy are not causal therapies and often have severe side effects. As their replacement, we should step by step introduce detoxification treatment in connection with a holistic-medical approach, as is to some extent already practiced by naturopaths.

Environmental-medical diagnostics and treatment (environmental monitoring, biomonitoring) must include the following: dental restoration, detoxification, avoiding exposure to harmful substances, as well as better psychological patient counselling. Hence we should demand from our politicians and health insurance companies to reimburse the costs of examinations and treatment of chronic diseases and cancer, if their cause could not be found with the help of today's conventional medical examinations. This procedure is overall cheaper and carries much less side effects for the patient than painful radiation treatments and chemotherapy, both of which also are very expensive.

The successes gained by detoxification treatment have been substantiated by the Essen central office for documentation of naturopathic treatments (Zentrale zur Dokumentation der Naturheilverfahren in Essen), in cooperation with the health insurance companies, from 1995 to 2002. What applies to chronic diseases should also be possible for malignant tumors on the precondition that professional detoxification is carried out.

An important message pertaining to therapeutic treatment
Since we continue to be exposed to harmful substances, regular environmental-medical checkups and further detoxification courses are necessary. Noxa with a half-life[38] of up to 20 years cannot be removed in one single detoxification course.

In his article "The Turning Point in the War Against Cancer", U. Bahnsen (2.) formulated it as follows: "In no way is cancer always a matter of fate. Every second patient could have remained healthy, and in Germany 70,000 tumor patients each year die a death that could have been easily prevented." Based on the DPA press release from December 7, 2012, these numbers may be even higher.

From our discussion of cancer we can see that with regard to environmental medicine, conventional medicine lags behind the current state of science to a considerable extent. It is possible in the future to perform environmental-medical examinations of patients with diseases of unknown genesis and treat them appropriately, numerous other malignant tumors can be prevented. I estimate that we can cure, or at least prevent in the future, easily 80 % of chronic diseases and cancers.

---

38 A substance's half-life is the time it takes for half of the amount of a certain substance to be metabolized and excreted from the body. It is easy for anyone to understand that under normal circumstances mercury, which has an 18-year half-life, will never be completely excreted in one lifetime. Hence, it can be transferred from pregnant women to their descendants.

13.3. Legal assessment

In both of the first two cases, the number of rejected claims for damages for patients who presented environmental-medical case evidence was at almost 95 %, even though from a medical point of view a 100 %-certain correlation had been established. This was not sufficient from a legal point of view. The judge almost always followed the expert reports presented by the public and private health insurance companies or trade associations, even though these were frequently scientifically unsound, incomplete, and riddled with errors. Many times I got the impression that the content of the report contained a lot of pre-programmed boilerplate—in any case, I saw the exact same jargon terms and psychiatric diagnoses reused time and again.

This type of environmental-medical jurisprudence is clearly reflected in the two examples provided by the case of the child with a brain tumor and the case of B., who had to fight for 12 years to get his psychiatric disease recognized as having been caused by toxic brain damage. It seems that in this domain of medicine the legal system is dictated by political dictates. Many colleagues report similar experiences in expert witness procedures that they encountered.

The prescribed rule for the judge is to at first (and at second) reject any claims for damages, if possible. Especially if the claims are made against public authorities and/or public health insurance companies, pension funds, as well as trade unions, the chances of success are almost zero, as we have seen in Matthias' case.

In that trial, the city authorities were acquitted, even though they had never performed regular measurements of PCB levels—something which they are in fact legally obliged to do. Twice, tests for harmful substances were carried out, after pressure exerted by one of the teachers. According to Matthias' mother, the results of these tests have never been made public—something which is rather unusual!

The way these cases proceed for the defense lawyer and the accused authorities or insurance companies is always the same. The latter select several expert witnesses. Usually this includes university professors, a psychiatrist, a neurologist, and a psychologist or psychosomatic specialist. The patient has no such opportunity: they usually cannot find other medical expert witnesses—at least I never have seen it happen that the defense lawyer invited his client to obtain expert testimony from a second or even third expert with qualifications in environmental medicine (e.g., a toxicologist or epidemiologist), or themselves took steps to obtain one.

> *In practice, each patient with environmental damage is happy when they find someone willing to prepare an expert witness report at all.*

In the progress report prepared by W. Krahn-Zembol (67.), which contains a list of publications ordered by theme and keyword, you can find an overview over jurisprudence and legal developments in Germany, in the field of environmental medicine and toxicology.

In light of the steadily rising number of patients with environmental damage, the lack of qualified environmental physicians and hence of independent environmental-medical expert reports will aggravate the current situation even further, from a medical and legal point of view. This applies to both patients and healthcare practitioners.

Hence, the only cases that stand any chance of success are those that make it to the federal social court (Bundessozialgericht), which may then go on to try and reach the Federal Constitutional Court (Bundesverfassungsgericht) as a precedent. At least here, the laws written in the federal code of social law still apply. The medical significance of this fact testifies to the necessity of a change in thinking.

> *Every doctor and every naturopath should also receive environmental-medical training.*

As an immediate consequence, jurisprudence too should adapt to the new facts. With regard to environmental medicine, the way jurisprudence is established somehow reminds me of a passage from Aino Kuusinen's memoirs (69.). Ms. Kuusinen, a Finnish communist activist, had asked W. Leonhard to publish her book only after her death. Under Stalin's rule, she was arrested by the Russian secret service. After her arrest, her interrogating officer, a high-ranking official, asked her: "Do you know the difference between civil justice and our legal system?" Of course, Mrs. Kuusinen didn't know the answer, which the official then provided: "In civil justice, the public prosecutor must prove that the defendant is guilty. In our legal system, it is expected that the defendant proves that they are innocent."

Even scientifically sound expert reports that show unambiguous correlations are ignored, as shown in Matthias' example. Such verdicts are painful defeats for both patient and doctor, because they cannot understand how something like this can happen in a democratic state with a strong sense of right and wrong and large ethical and moral responsibility.

Who then is to blame for such a wrongful verdict? Here, I will not try to answer this question. I can only describe a possible history for such decisions, which is as follows: in industry, manufacturers develop chemicals and medical drugs and products. If these substances cause any consequential damage, the manufacturers are usually not made liable, as can be seen in the case of thalidomide. The same applies to nuclear power plants and the consequential damage they cause. Children in the vicinity of a plant, who suffer from leukemia, have hardly any chance of compensation or support from the plant's operators. To my knowledge the manufacturers don't have any fund from which they pay compensation for damages.

Both big industry itself and the politically responsible parties are by all means aware of the danger posed by many chemicals and metals. One clearly visible symptom of this is provided by the numerous toxic waste dumps, scattered all throughout the country and marked with yellow-black warning signs that say "Toxic waste deposit site! Do not enter!". These sites all have one giant defect: they frequently are not sufficiently isolated from the soil underneath. The consequences are contamination of groundwater and soil, as well as air pollution. The question is: where to go with this waste?

We can only solve this problem if, in the future, we produce less poison.

In the current legal landscape, a single person has hardly any chance to win a court case. As a consequence of these wrongful verdicts, special interest groups have been founded in the vicinity of atomic power plants, toxic waste disposal sites, hydrofracking drilling sites, and mobile communications masts. These groups argue that radiation, contaminated groundwater, air, and vegetation, have caused an increased number of diseases, in particular cancer. As a consequence, several court cases have been started by now, which may run for years and even decades. These groups rightly demand closing nuclear power plants, stopping drilling for hydrofracking, closing dangerous toxic waste disposal sites, and the deactivation of mobile communications masts that are located in the vicinity of residential areas.

However, for understandable reasons, politicians and other responsible parties do not see any dangers to our health. They trust the purportedly scientifically sound results that they are presented with by the companies. There are no neutral assessments of these results—who would pay for that, with the health insurance companies being out of money?

This again brings us back to our starting point:

Today, economic interests take precedence over our health and we are forgetting or ignoring our moral and ethical values.

The burden of proving their disease being caused by exactly these chemicals or medical products is lying on the shoulders of the weakest member in the chain of evidence. Proof must be 100 % sound both in a medical and in a legal sense, which is hardly possible in practice.

The numerous industrial and nuclear accidents time and again lead us humans towards the abyss the same way that it does to animals and plants. Each of our lives, as well as that of or progeny, irrevocably depends on the flourishing of our animals and plants. We must take care that the snowflake that has already become a large snowball does not turn into an avalanche. Legally, this problem can only be solved by reversing the burden of proof. In the future, manufacturers should be required to prove the safety of all their products.

13.4. Moral and ethical assessment

If one assembles the individual pieces of the mosaic, one has to conclude that a change in thinking is not only necessary from a medical, but also from a moral and ethical point of view. Here us doctors in particular must take our responsibility. Hence, I see myself primarily as general and holistic physician, who tries to continuously keep in mind not only the whole field of environmental and general medicine, but also the social and sociopolitical situation.

I am not alone in this attitude. During some of his Whitsuntide sermons, pope Francis pointed out that industrialization is there for the people and not vice versa. Clearly, the catholic church is aware of its moral and ethical mission. Nevertheless, this realization is clearly absent from the minds of some politicians and multinational companies. I fail to understand this, because it is exactly this way of thinking that emerges from our federal code of social law, which is already as much as 60 years old. This fact should give us courage and hope.

As early as in §1 of our federal code of social law, it is stated that "the laws in this book serve to create social benefits, including social and educational aid, to help realize social justice and social security. They must contribute to secure a life fit for human beings, create equal opportunities for free personal development—in particular for young people—enable self-sustenance through a freely-chosen profession, and—also by supporting self-help—prevent or compensate for the special burdens that our lives place on our shoulders."

When it comes to victims of environmental health damage, these four points are, by and large, not satisfied. Because legal proceedings, as mentioned earlier, often drag along for many years, the people involved in the case are mentally and physically taxed so much that they become even more ill. These processes drop them into a depression, which makes them even more dependent on medical drugs that they don't tolerate. The story of my patient who suffered a stroke after the verdict in her case was announced is only one example among many.

Our natural-scientifically oriented contemporary medicine has achieved many excellent successes, in particular in areas such as acute medical care, the fight against infections diseases such as poliomyelitis, tetanus, rabies, and the plague. The same applies to all specialties within surgery and obstetrics, neonatal care, transplantation medicine, neurosurgery, etc. This fact must most empathically be mentioned here.

Regrettably, the same is not true for chronic diseases and cancer. Here, conventional medicine clearly is at a dead end, because it excludes the complete field of environmental medicine. There is hardly any research on the causes of diseases and only symptoms are treated. Patients fall victim to statistics and norm value and often undergo unnecessary surgery, as we can hear and read in the media time and again.

The magic word that health insurance companies have introduced is "flat rate payment". In more casual terms, one might say that "we want to be a few steps ahead of the doctors"—this is the only way a hospital can operate in an economically viable way. Under these circumstances, how is a doctor supposed to fulfill his Hippocratic oath if he has to treat ever more patients in ever shorter time? There is no time for true healing and prevention, as well as for environmental medicine, which itself represents preventions. They answer that "there is no headroom for that".

If we here use the Chinese proverb as a measure of the quality of the job we do as doctors, then we must shamefully admit that we are the "small-time doctors", who only treat existing diseases or their symptoms.

There are two reasons for this deterioration of medical care.

1. The consciously intended, ever more strict confinement of our medical action. The keywords here are inflated bureaucracy and quality control. Under these well-intended codenames, external parties are tightening the belt around our freedom of action ever more strongly. After all, how could anybody be against quality control? Regrettably these controlling procedures take a lot of time away from medically more urgent tasks, such as patient interviews, nursing, and for necessary regular house calls by the family physician or medical specialist. Another example is provided by the significance of preparing a comprehensive medical history, which I have pointed out time and again in this book. In retirement homes these problems are even more pronounced, as I am experiencing each day with my mother-in-law, who lives in such a facility. Even though the nursing staff works themselves to the point of collapse, there is no time left for individualized care. The latter is, however, absolutely necessary and is much more important than some work protocol. If there were extra staff available for these administrative tasks, it would be alright. The retirement home that wins at bookkeeping receives the most praise. A lot more negative criticism is heaped upon any facility that invests a lot of time in nursing and other care but takes less time for keeping written records.

2. In the past, we doctors haven't had sufficient civil courage and didn't sufficiently fought back against the ever stricter constraints and controls that are required by administrative departments, health insurance companies, and politics. In our defense, I should however say that politics, infighting, and strikes are not our thing. First and foremost it was the chairperson of the Association of Statutory Health Insurance Physicians (Kassenärztliche Vereinigung) and the presidents of the regional medical associations who failed, since they are professional politicians, who were elected to provide relief for us doctors. As employees of a public agency

(authorities) and in light of their well-endowed contract, they may have silently accepted many things that they shouldn't have.

I cannot forget the question as to how to recognize a patient with environmental damage, which was posed by a chairperson of the KV Dortmund. We cannot reproach our colleagues who work in private practices and hospitals. We have had to look on as, during the past 25 years, our medical science has been going further downhill. Despite medical and technological progress, the number of so-called diseases of affluence has increased immensely—we might even say that this is precisely due to these developments. Today we believe that technology can replace human beings. In this, morality and ethics are left behind.

Since 2011 there has been a small glimmer of light, after we as environmental physicians gained rear cover from a resolution made by the Council of Europe in 27.05.2011. In this resolution, the Parliamentary Assembly of all member states demands to remove as many toxic metals as possible from our human environment and to prevent their bioaccumulation in nature as well as their entrance into the food chain and the human body.

The Council of Europe calls out to reduce our environment's exposure to heavy metals.

I have focussed on describing metals and the health damage risks that they pose. I consider this to be the biggest danger. This may however change soon. Each year, thousands of tons of chemicals are produced and enter trade—most of them without having undergone any testing for tolerance. Furthermore, during the past 20 years, health damage due to the explosive increase in electromagnetic radiation and its mutual interaction with other noxa has increased dramatically.

In 2006/2007, the EU presented a draft for a reorganization of the regulations that cover chemicals. This program was called REACH and is intended to regulate registration, evaluation, and authorisation of chemicals. In the future, and under the motto "no data, no market", only those substances may be traded for which there exists sufficient data, the type and area of coverage of which is determined primarily by the intended production level. This sounds good. Clearly, the EU commission is not only aware of its medical but also of its moral and ethical responsibility. Regrettably, it has not yet been able to effect much change, particularly because non-European countries are not bound by these regulations.

Nevertheless, REACH is a small step in the right direction. It provides us consumers at the bottom of the ladder with some hope that the danger from harmful substances from the environment now is also recognized and dealt with "at the top". This will certainly increase the status of environmental medicine within the complex whole of our social and political life in the future. The at the time "tiny cog" that is environmental medicine may soon become a big wheel in the mechanism of our healthcare system as a whole.

The industrialization process, and the social and political developments that go hand in hand with it, has been waltzing over us for more than 200 years. It has created our welfare state and stamped its mark on our thinking and action. Our task is to halt the negative developments caused by these processes. We now must look forward and make a thoroughfare out of the small step that is REACH, which will lead into the whole world. Only if we are all convinced that something must change and everybody makes their contribution then all will change for the better.

I recently held a lecture on the topic of "Metals: the Hidden Causes of Chronic Diseases", in a small town. By sheer chance, during my talk my eyes fell on a poster on which was written, amongst other things, "fibromyalgia is incurable".

I was surprised. I interrupted my talk and pointed the audience's attention to this line. With remarks like this, the group of fibromyalgia patients degrades itself to the status of "pity group". The question emerges as to how a member of this group can ever regain their health if they read lines like that during each and every meeting. By now, the line has been changed to "fibromyalgia is curable".

With this, we leave the sociopolitical level and return to our starting point, which is the situation of the affected patients. A precondition for curing an individual patient is that they believe in it themselves, and spend efforts towards reaching that goal. The examples that I have presented show that even patients who are apparently "incurable" and who have been given up on by conventional medicine have regained their health thanks to their own initiative and that of their fellow humans (parents, spouses, friends, family physician). I would like to encourage all of my readers to follow these examples.

This book should serve as more than just a report of my experiences, more than a summary of scientific studies and their results. It is intended as an analysis of a negative medical development, which is already wreaking havoc also in the world of animals and plants.

At the same time, I want to offer the numerous people with environmental damage a new perspective on life and boost their self-confidence. I want to prevent that contemporary medicine, which has been so outstanding in many areas, reaches a dead end when it comes to chronic diseases and cancer—a dead end where there is no longer any way out for many people.

A fundamental question: who actually is to blame for the loss of moral and ethical values in the field of environmental medicine and of medicine per se? This loss can also be observed in all areas of politics, economics, finances, sports, and even in the churches. The answer can only be that we are all jointly responsible. Therefore, we are all summoned to turn on our heels and to pull the wagons that carry the topics of "medicine" and "declining morals" out of the quagmire, before everything sinks below the surface.

14. Excerpt from an open letter by Prof. Wassermann

The field of environmental medicine can find support in an open letter to former prime minister Heide Simonis, written by Prof. O. Wassermann, from Juli 20, 2000 (120.). The letter was written against the background of the closing of numerous institutes for toxicology or their incorporation within institutes for pharmacology and hygiene.

In his capacity as coordinator of the "Coordinated action against dangers posed by BAYER", Prof. O. Wassermann wrote: "This coordinated action against the dangers posed by BAYER (Coordination gegen BAYER-Gefahren—CBG), carried by more than 1,000 supporters, demands appropriate facilities and political backing for its institute. For 20 years, CBG has been campaigning for a world free of toxins. Frequently, the association takes up results obtained by critical toxicologists and chemists. As Philipp Mimkes, director of the CBG, puts it: 'Every day we are confronted with new chemicals. Thousands of them have never been tested for their toxic potential.' Mimkes laments the fact that most german experts are dependant on orders issued by the chemical industry. To date, the University of Kiel Institute of Toxicology has been one of the laudable exceptions. Only autonomous research, with proper facilities at its disposal, can promote public health. Where appreciation for toxicology is significantly on the rise in other European countries, seven out of eleven professorships were shut down in Germany. Experts fear a deterioration in the safety of medical drugs and a decrease in occupational safety."

There is no better way to describe the situation that environmental medicine finds itself in, together with the lurking dangers posed by potential hidden toxins. Since the situation has deteriorated rather than improved, despite appeals such as the one above and others like it, that what I said at the outset holds: regrettably, within the foreseeable future there is no real help to be expected from politics and commerce.

Finally, as a holistic physician I would like to shout out to my readers and all the affected patients who, despite all efforts still haven't found any critical medical help:

Ask yourselves: am I suffering from an environmental disease?

If you answer 'yes' to this question then, to close circle, my advice to you is:

Take your health into your own hands!

My experiences with patients who suffer from an environmental disease are based in the past. I have tried to look back into history and found that we can learn a lot from it. For this reason, I adopted those insights in the present day.

I have analyzed the present day and found that despite all the dangers life is, as it was before, a miracle and the things that we experience day after day are miraculous. That what we know about life is only a large "Nothing"—but we behave as if thanks to our natural-scientific insights we have life itself under control.

I have tried to look into the future and found that we are at a crossroads. I am very sure that everyone will find their own way, which is the right path for themselves in order to be in harmony with nature. If this happens, the question I asked at the outset will be superfluous.

15. My wishlist for the future

My wishlist for contemporary medicine

- May contemporary medicine take up the experience and achievements gained by the advanced civilizations of the past.
- May it integrate the advantages of our contemporary so-called alternative medicine into its therapeutic concept. We need medicine without side-effects.
- May environmental medicine take its due place in contemporary medicine and hence in our public health system. It represents, as its name already indicates, a healthy environment.
- May environmental medicine become a teaching and research subject in all medical academies and universities.
- May detoxification therapy, as part of a holistic therapeutic approach, replace chemotherapy and radiation treatment.

My wishlist for our jurisprudence

- May the burden of evidence in legal proceedings be reversed as soon as possible. It cannot be that a severely diseased patient must prove that their disease was caused by harmful substances.
- May §1 of the social security statute book not be only printed on paper, but also practiced in daily life (see p. 283!).

My wishlist for our politicians, industrial companies, all responsible parties, all doctors, and all healers as regards morals and ethics.

- May the big industrial companies always place the good of mankind first in all their research and other endeavours.
- May industry and economy once again place themselves at the service of mankind.
- May politicians, ministers, and representatives apply their thinking and action in concordance with the name of their occupation (minister derives from the latin ministrare = to serve).
- May all healers, conventional or alternative, hunt down the causes of diseases and strive towards real prevention instead of just remedying

symptoms. We all would like to be, or at least become, one of "the best" doctors, right?

15.1. A greeting from the "Natur & Heilen" (90.) publishing company

I would like to conclude with a greeting that I found in volume 4/2012 issue 3 of the "Natur & Heilen" publishing company: "That which was and is self-evident for many primitive people, namely that trees and all other plants have a soul, is increasingly being confirmed by modern science. We are directly connected with this, our environment. The more conscious we are of this fact, the more we can open up ourselves to the energetic level of the floral world and feel its healing bond.

Wangari Maathai, a nobel laureate who won the Nobel Peace Prize and through her commitment caused millions of trees to be planted in Africa, goes even further. She invites us to regard the environment as sacred, because its destruction implies the destruction of life itself.

This becomes clear to us each year, when we look on in wonder when nature wakes up. We would like to discard all our winter sleepiness so that we are well-prepared for this new cycle of life."

During my stay in Africa, I experienced the knowledge and belief that emanates from this greeting every single day whenever I was interacting with other people. We can learn a lot from them too.

We are living at the beginning of a new chapter in medical history. Since industry, economics, and science penetrate our health as human beings to an ever larger extent, these three sectors must also do more research on the risks that go with their actions, in order to protect the health of humanity.

15.2. An appeal, and a "Thank you!"

This book is aimed at all laymen, in particular at victims of environmental damage. It intends to give them courage and hope for being able to make a large contribution to maintaining their health themselves and, in cases of a chronic disease, to be able to conquer it. The same is valid for cancer.

This book is aimed at all naturopaths, homeopaths, acupuncturists, and holistic physicians. I am feel connected with them and am thankful to them, because they have further cultivated the art of healing with medication without side effects, as taught by our forefathers, even in this age of industrialization.

My thanks goes out to all colleagues, including those at dentistry faculties, and to all who work in healthcare, and I call out to them to work together with us. Through their help, enormous progress has been made especially in the field of emergency medicine, surgery, and the study of infections diseases. However, in order to master the tasks that are ahead of us, I also call them to practice a more holistic type of medicine while never neglecting their specialization. Only then will environmental medicine get the attention and recognition that is its due. Your patients will thank you for it.

Furthermore, this book is an appeal to all responsible parties within our healthcare system to recognize environmental medicine as a subject of research and teaching, and to carry the costs of diagnostic examinations and therapeutic treatments. This is necessary from a medical, legal, moral, and ethical point of view, because that way much suffering can be prevented. If we follow the mainstream, our healthcare system soon will no longer be affordable for everyone.

The comprehensive studies that are documented in the literature, written by the most excellent researchers, together with my own studies in my capacity as family and environmental physician, carried out for more than 25 years (1987 to 2012), provide sufficient reason for being hopeful that we can cure—or at least prevent—many environmental diseases, including cancer. They give us courage to proceed along our path. Achieving this requires the cooperation of every single individual.

16. A few words for you to take home with you (Prof. Dr. med. Frentzel-Beyme)

Do I have an environmental disease?

So many people ask themselves this question, that the topic of discussing these daily observations from medical practices merits an extraordinary amount of attention.

After the environmental catastrophes in Japan (Minamata, Yusho, etc.), both government and industry felt obliged to establish a university for occupational and environmental diseases in Kitakyushu. This university, which contains five separate institutes, was founded in 1978 as a compensation for the various environmental diseases caused by those catastrophes that expose a defenseless population to toxic substances. This institute of higher learning educates doctors and medical staff to consciously monitor previously unknown, newly developed clinical pictures and sponsors scientific inquiry. Through all of his work, the founder of this University of Occupational and Environmental Health, Prof. Kenzaburo Tsuchiya, has acknowledged the in his opinion important role of the so-called "astute physician". His motivation is that, as the most important initiator of the healing process, the accomplishments of this type of doctor must be classified as of even higher importance than epidemiological scientific studies. Only after such an astute physician has hypothesized a causal connection between source of a disease and the disease itself, and in the presence of a well-argued suspected cause, only then will the necessary scientific studies be plannable. If this is all in place, these studies will then indeed be funded, if political will towards proactive preventive environmental medicine is there. In this respect the progress made in Japan over the past almost 35 years are exemplary, even though they are not extraordinary. In Germany, however, we regrettably lack such a focussed approach.

During his time as a medical professional, and in his capacity as astute clinical doctor, Walter Wortberg has gathered many special impressions. He tapped into this wealth of experience in order to pursue his suspicions in a much more systematic way than many of his colleagues have done. Many of them make similar observations, which they store in their bank of experience, and go on to pay more conscious attention whenever they encounter similar phenomena. However, due to work overload, they aren't able to research the accumulating causal roots of these diseases which, had they had the time, could have been prevented. Walter Wortberg was not content to merely observe and wanted to join the quality of his observations with quantitative results that are, if possible, convincing beyond any doubt even in the presence of the by now common repression of any disturbing knowledge. By his sheer engagement, without any external support and in parallel with his taxing everyday work, he prepared documentation that allows him to unite his observations and back them up with numbers that make it highly unlikely that all the cases he describes are nothing more than coincidence.

The results of his efforts are condensed, in the form of descriptions and pictures, into the contents of the present book. This work also shows a doctor's everyday work as reflected in the affected patients – a doctor to whom worries about avoidable exposure are a continuous source of concern.

Prof. Dr. med. Rainer Frentzel-Beyme
Environmental physician, epidemiologist, and co-founder and -publisher of the magazine "environment – medicine – society" ("umwelt – medizin – gesellschaft")

17. Appendix

17.1. Some words of thanks

I would like to thank all colleagues, naturopaths, dentists, or rather all persons listed in the bibliography. During my 25-year-long travels through the world of environmental medicine—through the history of medicine—their articles and books have time and again revitalized me and given me courage so that I stayed on the right track during my studies. This path did not always go straight ahead and was strewn with many obstacles. From the beginning, I walked a narrow line. It was a challenge, since it was necessary to explore virgin soil.

I would like to thank the nuns of the Franciscan nunnery in Olpe, who for a week once or twice a year have treated me almost as their child and maintained this for many years. Whenever I was with them, I drew strength and found the peace to continue writing my book.

My thanks go out to all my friends who proofread various drafts of the manuscript: Johannes Broxtermann, Ricca and Hilmar Edmondson, Bernd and Johanna Henrichs, Lothar and Barbara Klemisch, Matthias Wagner, and my cousin Elisabeth Freitag. My regular conversation partners: Monika Frielinghaus and Gisela Grote, with whom I've been in a continuous exchange of ideas during all those years. They have always encouraged me to hazard the balancing act of writing a book a) for laymen who are in need of enlightenment and b) that raises scientific claims. The truth is always simple and laymen have the right to learn it.

My friends from Ludenscheid Franz and Roswitha Bierer, and Helmut and Dorothee Mlitz. Between us, environmental medicine never ceased to be a topic, whether at home or during one of our numerous mutual walks. Laymen downright thirst for information and truth, of which they are given ever less from an official side.

I would like to thank the naturopath Anita Duda: in my capacity as holistic physician, I have worked closely with her. She reinforced my belief that natural medicine as a type of medicine without side-effects must be treated on equal footing with conventional medicine. In the future, every physician should master the basics of both these therapeutical approaches.

My special thanks goes out to my family, which had to suffer my absence not only during everyday life, but also during practically every single holiday, due to my continuing retreat for research.

Maybe I will be able to make up for it to my grandchildren. I am thankful to my grownup sons Daniel and Hendrik for their numerous tips and advice. Without the help and understanding of my wife Monika, who proofread every word I wrote, this book would not have existed.

I also wish to thank Ms. Zimmermann and Ms. Maiworm from M.A.M. Maiworm GmbH, who put the finishing touches on this book.

17.2. Explanations and definitions

ADI: Acceptable daily intake.
Alloys: mixtures composed of different metals.
Biomonitoring: measurement of harmful substances in blood, urine, mother milk, hairs, teeth, and tissue.
Black Serpent Stones: a stone known in Latin America, Africa, and Asia for its healing properties. It is placed on snake bite wounds and other wounds caused by the sting of poisonous animals to suck the poison out of the wound, so to speak (adsorption).
Chelating Agents: from the greek root "chele" = lobster's pinchers; substances that chemically clasp and surround other substances. This process creates a chemical bond with some component of the chelating agent clasping a metal.
Coherency: defined to be the sum of all correlating properties between wave quantities. Often, coherency indicates a mutual genesis. All physical waves, such as light rays, radar waves, sound waves, or fluid waves can be coherent with other waves in some sense, or there can be coherency between respective partial sub-waves.
Creatinine: a metabolite excreted by the kidneys. Its levels are measured to check kidney function.
Creatinine clearance: clearance in this sense means elimination, removal, purification, or clarification. It designates the amount of plasma from which a certain amount of substance (e.g., creatinine) is removed per unit of time. Different substances are excreted at different speeds. Tabulating the creatinine measurements is outside of the scope of this work and for that reason

I have not listed them. However, there exists a correlation between heavy metal exposure and creatinine clearance values.

Deletion: genetic damage.

Dental dam: the oral cavity and all teeth not affected by the dental work are covered with a plastic membrane which prevents mercury-containing fine dust from entering the oral cavity, from which it might enter the lungs or the brain directly.

Dental restoration material: dental filling material, metallic materials, dental alloys.

Diuresis: Excretion in the urine

Environmental monitoring: measurement of levels of harmful substances in water, soil, foodstuffs, domestic objects, and building materials.

Epidemiological studies: In medical science, epidemiologic studies are conducted to determine the frequency, distribution, and causes of diseases. When causal proof of diseases is concerned, we speak of analytical epidemiology.

Exposure: any measured value that is above the reference value is designated as exposure.

Exposure to heavy metals: determined if lymphocyte transformation test (LTT) shows hypersensitivity in the blood. The measured parameter is the Type IV cellular hypersensitivity. In the text, we discuss the stimulation index (SI). An SI > 3 corresponds to a more than threefold activation compared to baseline and shows the existence of allergen-specific T cells in the patient's blood (positive result), i.e., cellular hypersensitivity. An SI < 2 counts as a definite negative result. An SI between 2 and 3 should be considered borderline (weak, questionable hypersensitivity) and monitoring the patient should be considered.

Fibroblasts: connective tissue cells that play an important role in the synthesis of intercellular material.

Half-life: a substance's half-life is the time it takes for half of the amount of a certain substance to be metabolized and excreted from the body. It is easy for anyone to understand that under normal circumstances mercury, which has an 18-year half-life, will never be completely excreted in one lifetime. Hence, it can be transferred from pregnant women to their descendants.

In vitro: from the Latin "in glass"; designates organic processes that take place outside of a living organism—this in contrast to processes that occur in a living body. The natural sciences use in vitro experiments that can be conducted in an artificial environment outside of the living organism, e.g., in a test tube.

In vivo: from the Latin "in the living"; studies on living organisms.

Load factor: the quotient of the average of the measured value divided by the reference value.

Metal-symptogram: this concept says that it is possible to diagnose exposure to metals on the basis of symptoms.

Muscular hypotonia: reduced muscular tension when in motion, caused by CNS disorders and disorders in the spinal cord.

Oncopheresis: apheresis performed with a specialized double diaphragm filter (135.). Oncopheresis is a new concept designating therapeutic use of apheresis (see p. 211 for the definition of apheresis!). I only use it when I prescribe apheresis for patients with a malignant tumor and simultaneous exposure to harmful substances.

PCR-testing: during a polymerase chain reaction (PCR) certain nucleic acid fragments, e.g., from viruses, are duplicated and visualized by staining after separation through gel electrophoresis*. The advantages of this method are that minute amounts of DNA can be measured with high sensitivity and with rapid results, as well as its capability to show presence of viruses that cannot be grown in cell cultures.

Pharmacogenetic testing: this is a genetic test for tolerability where enzymes responsible for metabolizing (and hence excreting) medical drugs and their metabolites are tested. In addition, it is possible to test medical drugs for insufferable side-effects (sensitization). Today this is necessary, because ever more people exhibit allergic reactions and the number of medical drugs with side-effects is continuously increasing. The stronger the drug, the more frequent the occurrence of hypersensitivity.

Polymorphism: variations in genetic frequency; changes to the DNA sequence.

RAC (reflex auriculo cardiac) after Nogier: a diagnostic method that has its roots in ear acupuncture (auricular acupuncture). By taking the patient's pulse in a particular way, the correct acupuncture points can be found. Ad-

---

*) Gel electrophoresis is method from analytical chemistry. It is used in chemistry and molecular biology to separate different types of molecules

ditionally, allergies, toxic exposures, and drug tolerability can be tested. Dr. Paul Nogier was a French physician from Lyon, who discovered this reflex in 1966.

Reference and norm values: no reference values for metal levels in tissue exist. Hence, I compared the values found to the reference values in empty-stomach urine to have at least some indicators for interpreting the results of the tests.

Screening: a systematic testing procedure that is used to identify certain properties of the tested subjects within a well-defined area of testing—usually a large number of samples or test persons.

Tin-symptogram: lists symptoms that indicate exposure to tin or organotin compounds. It applies particularly to children.

17.3. Addresses: associations, laboratory institutes, clinics

Research labs
Institut für Medizinische Diagnostik Berlin-Potsdam MVZ GbR
[Institute for medical diagnostics in Berlin-Potsdam]
Nicolaistr. 22
12247 Berlin-Steglitz
Tel.: 030 / 770010
Fax: 030 / 77001332
info@imd-berlin.de
www.imd-berlin.de

IPgD Institut für Pharmakogenetik und genetische Disposition
[Institute for pharmacogenetics and genetic disposition]
Ostpassage 9
30853 Langenhagen
Tel.: 0511 / 2030448
Fax: 0511 / 2030447
es@ipgd.org
www.ipgd.org

Medizinisches Labor Bremen
[Bremen medical laboratory]
Haferwende 12
28357 Bremen
Tel.: 0421 / 20720
Fax: 0421 / 2072167
info@mlhb.de
www.mlhb.de

Dr. Lorenz – Institut für Innenraumdiagnostik
[Dr. Lorenz – Institute for interior diagnostics]
Marconistr. 23
40589 Düsseldorf
Tel.: 0211 / 99958160
Fax: 0211 / 99958177
infid@infid.de
www.infid.de
Labor Dr. Fenner und Kollegen

Medizinisches Versorgungszentrum für Labormedizin und Humangenetik GmbH
[Medical therapeutic center for laboratory medicine and human genetics]
Bergstr. 14
20095 Hamburg
Tel.: 040 / 309550
Fax: 040 / 3095513
fennerlabor@fennerlabor.de
www.fennerlabor.de

Environmental clinic
Spezialklinik Neukirchen GmbH & Co. KG
[Neukirchen specialized clinic]
Krankenhausstr. 9
93453 Neukirchen b. Hl. Blut
Tel.: 09947 / 280
Fax: 09947 / 28109
info@spezialklinik-neukirchen.de
www.spezialklinik-neukirchen.de

Environmental-medical organizations
Deutscher Berufsverband der Umweltmediziner (dbu) e. V.
[German environmental physician's professional organization]
Siemensstr. 26a
12247 Berlin
Tel.: 03077 / 15484
Fax: 03077 / 15484
dbu@dbu-online.de
www.dbu-online.de

Geschäftsstelle der Deutschen Gesellschaft für Umwelt – ZahnMedizin e. V.
[Office of the German society for environmental dentistry]
Siemensstr. 26a
12247 Berlin
Tel.: 030 / 76904520
Fax: 030 / 76904522
info@deguz.de
www.deguz.de

European Academy for Environmental Medicine e. V. (EUROPAEM)
Trier Straße 44
54411 Hermeskeil
Tel.: +49 (0)6503 98 10880
Fax: +49 (0)6503 98 10881
congress@europaem.de
www.europaem.de

IGUMED Geschäftsstelle
[Office of the IGUMED]
c/o Labor Dr. Fenner & Kollegen
Bergstr. 14
20095 Hamburg
Tel.: 040 / 30955494
Fax: 040 / 30955495
info@igumed.de
www.igumed.de

Ökologischer Ärztebund e. V.
Bundesgeschäftsstelle
[Federal office of the ecological physicians society]
Frielinger Str. 31
28215 Bremen
Tel.: 0421 / 4984251
Fax: 0421 / 4984252
oekologischer.aerztebund@t-online.de
www.oekologischer-aerztebund.de

Environmental-medical self-help groups
Verein zur Hilfe umweltbedingt Erkrankter e. V. (VHUE e. V.)
[Society for the support of patients with environmental diseases]
Monika Frielinghaus
Hallstattstr. 2a
91077 Neunkirchen a. Br.
Tel.: 09134 / 909008
Fax: 09134 / 707100
info@umweltbedingt-erkrankte.de
www.umweltbedingt-erkrankte.de

Selbsthilfeinitiative Umweltkranke Saterland – Sus
[Saterland self-help initiative for patients with environmental diseases]
Gartenring 2a
26683 Saterland
Tel.: 04492 / 913855
shi-umweltkranke@gmx.net

17.4. Bibliography

All cited literature is in German, unless indicated otherwise.

1. Altrock, Th.: Gefahren durch Dentalwerkstoffe und Umweltnoxen, Diagnostik und Therapie mittels Aurikulomedizin, Homöopathie und Bioresonanz, Hüthig Verlag, Heidelberg, 1997
2. Bahnsen, U.: Die Wende im Kampf gegen Krebs, Die Zeit Nr. 29/12, Die Zeit Verlagsgruppe Hamburg, 2007, pp. 33–36
3. Bahr, F.: Einführung in die wissenschaftliche Akupunktur, Ohr-, Schädel- und Körperakupunktur, MMV Medizin Verlag, München, 1996
4. Baudisch, Chr.; Prösch, J.; Panhans, H.: Polychlorierte Biphenyle in Fugendichtungsmassen ostdeutscher Plattenbauten, VDI Berichte 1122, VDI-Verlag GmbH, Düsseldorf, 1994, pp. 93–97
5. Bayerisches Staatsministerium für Gesundheit, Ernährung und Verbraucherschutz, Fachinformation Umwelt und Gesundheit, Zusammenfassung PCB, 2001, pp. 1–10
6. Beuth, J.: Krebs ganzheitlich behandeln; Trias Verlag im MVS, Medizinverlage Stuttgart GmbH & Co. KG; Stuttgart, 2007
7. Beyersmann, D.: Effects of carcinogenic metals on gene expression, Toxicology Letters, 127, Elsevier Science Ireland, 2002; pp. 63–68 [in English]
8. Bierbaumer, N.; Schmidt, R. F.: Lernen und Gedächtnis, Zelluläre Mechanismen, Neuro und Sinnesphysiologie, Springer Verlag, Berlin, Heidelberg, New York; 1998, pp. 409–415
9. Bierbaumer, N.; Schmidt, R. F.: Lernen und Gedächtnis, Neurophysiologie des Gedächtnisses – Gedächtnissysteme, Neuro- und Sinnesphysiologie, Springer Verlag, Berlin, Heidelberg, New York 1998, pp. 416–420
10. Bischof M.: Biophotonen, Das Licht in unseren Zellen, Zweitausendeins, 1999
11. Blech, J.: Giftkur ohne Nutzen, Der Spiegel Nr. 41, Hamburg, 2004, pp. 160–162

12. Blumer, W.; Raumann, R.; Reich, Th.: Krebsgefährdung durch Autoverkehr, Zeitschrift für Präventivmedizin, Revue de Médicine préventive 17,1972, pp. 157–161

13. Bonnet, E.: Sprechstunde Kinderkrankheiten, Hippokrates Verlag, Stuttgart, 1999

14. Brune, D.: Corrosion of amalgam, Scand. J. Dent. Res. 89, 6 Kropp Franz M. und Pantke, 1981, pp. 506–514 [in English]

15. BUND: Über 300 Schadstoffe in der Muttermilch, Endstation Mensch, Bund für Umwelt und Naturschutz Deutschland, Berlin, Juni 2005

16. Daunderer, M.: Chronische Intoxikationen, Diagnostik – Therapie – Prävention 1–3, Kompendium der Klinischen Toxikologie, part I volume 3, ecomed verlagsgesellschaft, Landsberg, 1994

17. Daunderer, M.: Gifte im Alltag, Der umfassende Ratgeber, C. H. Beck, München, 1995

18. Daunderer, M.: Klinische Toxikologie in der Zahnheilkunde, Diagnostik und Therapie, 3. supplement 7/96, ecomed verlagsgesellschaft, Landsberg, 1996

19. German Press Agency (DPA): Mehr Schwermetalle im Boden, press release in the "Westfälischen Rundschau" newspaper from 27. July 2011

20. German Press Agency (DPA): Mehr Schwermetalle in Industrieanlagenähe, press release in the "Lüdenscheider Nachrichten" newspaper from 27. July 2011

21. German Press Agency (DPA): Psychisch Kranke überrennen Praxen, press release (by Daniel Freudenreich) in the "Westfälische Rundschau" newspaper from 7. December 2012

22. German Press Agency (DPA): Tödliche Krebsleiden nehmen zu, press release in the "Lüdenscheider Nachrichten" newspaper from 7. December 2012

23. diagnose – FUNK: www.org / brennpunkt/ 20.01.2013, pp. 1–4

24. German Medical Journal (Dtsch. Ärzteblatt): Definition und Weiterbildungsinhalte des Bereiches Umweltmedizin nach der Weiterbildung-

sordnung für die Ärzte Bayerns vom 1. Oktober 1993, Deutscher Ärzteverlag Köln, 1993, 996, 93 (39), A 2460

25. German Science Foundation (Deutsche Forschungsgemeinschaft, DFG): Polychlorierte Biphenyle, Bestandsaufnahme über Analytik, Vorkommen, Kinetik und Toxikologie VCH, Verlagsgesellschaft mbH Weinheim, 1988

26. European Academy for Environmental Medicine (Europaem), Austrian Medical Association, Department of Environmental Medicine: Diagnostik umweltausgelöster Multisystemerkrankungen aus Sicht der Klinischen Umweltmedizin, Würzburg, 3. March 2012, Vienna

27. Fabig, K.R.: SPECT in der Umweltmedizin, umwelt-medizin-gesellschaft, 19/3/2006, UMG – Verlagsgesellschaft, Bremen, 2006

28. Fachinformation Umwelt und Gesundheit: Polychlorierte Biphenyle (PCB), Bavarian State Ministry for Regional Development and Environmental Issues, status on: June 1995, pp. 1–10

29. Faulstich, J.: Das Geheimnis der Heilung, Wie altes Wissen die Medizin verändert. Mens-Sana bei Knaur, Das Erste, Knauer Verlag, Munich 2010

30. Friberg L., Nordberg, G. F.; Vouk, V. B.: Handbook on the Toxicology of Metals, Volume I: General Aspects, Elsevier Science Publishers B. V. Amsterdam – New York – Oxford, 1986 [in English]

31. Friberg L.; Nordberg, G. F.; Vouk, V. B.: Handbook on the Toxicology of Metals, Volume II: Specific Metals, Elsevier Science Publishers B. V. Amsterdam – New York – Oxford, 1986 [in English]

32. Fritzsche, C.: Nebenwirkungen in Homöopathie und Allopathie, in H. Blog.: Homöopathie & Forschung, Berlin, 24. October 2007, pp. 1–6

33. Fritzsche, C.: Arzneimittel: Todesursache Nr. 3 in Deutschland, in the "Süddeutsche Zeitung" newspaper, Munich, 16.July 2008

34. Gerhard, I.; Runnebaum, B.: Grenzen der Hormonsubstitution bei Schadstoffbelastung und Fertilitätsstörungen, Zentralblatt für Gynäkologie 114, Georg Thieme Verlag Stuttgart, 1992, pp. 593–602

35. Gerhard, I.; Dermer, M.; Runnebaum, B.: Endokrine und Immunologische Veränderungen bei Frauen durch chronische Belastung mit Holzschutzmitteln. Lower Saxony state government, Hannover, 1992

36. Gerhard, I.; Runnebaum, B.: Schadstoffe und Fertilitätsstörungen. Genussgifte. Geburtshilfe und Frauenheilkunde 52, Georg Thieme Verlag, Stuttgart, 1992, pp. 509–515

37. Gerhard, I.; Runnebaum, B.: Schadstoffe und Fertilitätsstörungen, Schwermetalle und Mineralstoffe. Geburtshilfe Frauenheilkunde 52, Georg Thieme Verlag, Stuttgart, 1992, pp. 383–396

38. Gerhard, I.; Runnebaum, B.: Schadstoffe und Fertilitätsstörungen. Lösungsmittel, Pestizide. Geburtshilfe Frauenheilkunde 53, Georg Thieme Verlag, Stuttgart, 1993, pp. 147–160

39. Geuenich, K. und Hageman, W.: Kein Feuer ohne Rauch – Burnout erkennen, ansprechen und Hilfestellungen geben, umwelt-medizin-gesellschaft 4/ 2012, UMG – Verlagsgesellschaft, Bremen, pp. 227–231

40. Gleichmann, E.; Vohr, H.; Minschewa, W.; Stiller – Winkler, R.; Vogeler, S.: Zur Induktion von Autoimmunkrankheiten und anderen simulativen Immunopathien durch Quecksilber und andere Chemikalien. Umwelthygiene. Annual review (1986/1987), Volume 19, K. Beyen. Gesellschaft zur Förderung der Lufthygiene und Silikoseforschung e. V. Düsseldorf, 1987, pp. 270–319

41. Gienger, M.: Lexikon der Heilsteine, Fuldaer Verlagsanstalt GmbH, Fulda, 1997, p. 55

42. Grün, A.: 50 Engel für das Jahr, Ein Inspirationsbuch, Verlag Herder, Freiburg, 1997

43. Grün, A.: Was soll ich tun? Antworten auf Fragen, die das Leben stellt. Herausgegeben von Anton Lichtenauer, Verlag Herder Freiburg, Basel, Vienna, 2008

44. Grün, A.: Gelassen älter werden, Eine Lebenskunst für hier und jetzt, Herder Verlag, Freiburg, 2011

45. Grün, A: Trau deiner Kraft, Mutig durch die Krisen, Deutscher Taschenbuch Verlag, Munich, 2011

46. Haase, D. et al.: Increased risk for the therapy-associated hematologic malignancies in patients with carcinoma of the breast and combined homcygous gene of glutathione transferase M1 and T1, Leukemia Research 26, Publisher: Elsevier GB, Kidlingen, 2002, pp. 249–254 [in English]

47. Human Biomonitoring commission of the Federal Environment Agency, Reference values for PCB congeners Nr. 138, 153, 180 and their sum total in human blood. Federal magazine of health (Bundesgesundheitsbl.) Health science, Springer Verlag Heidelberg, 1998, pp. 41–416

48. Halbach, St. et al.: Entgegnung der Autoren des Materialbandes Amalgam im Spiegel kritischer Auseinandersetzungen, special volume: Institut der Deutschen Zahnärzte, Cologne, 2001

49. Halbach, St. et. al.: Amalgam im Spiegel kritischer Auseinandersetzungen, Interdisziplinäre Stellungnahme zum Kieler Amalgam Gutachten, "dental materials" series of the German Institute for Dentistry (Institut der Deutschen Zahnärzte), volume 20, Deutscher Ärzte-Verlag, Cologne, 1999

50. Hecht, K.: Patienteninformation zur prophylaktischen und basistherapeutischen Wirkung des siliziumdioxidreichen Klinoptilolith – Zeoliths, Heck Bio-Pharma Winterbach, 2011

51. Hennek, M.; Hennek, B.: Umweltbedingte Erkrankung EL/MCS, Gesund Wohnen & Leben Eigenverlag Umwelt- Beratung Würzburg, 2001

52. Hill, H. U.: Umweltschadstoffe und Neurodegenerative Erkrankungen des Gehirns (Demenzkrankheiten), Wie neurotoxische Langzeitwirkungen von Chemikalien zur Degeneration des Gehirns führen, Shaker Verlag, Aachen, 2012

53. Hirt, M.; M'Pia, B.: Natürliche Medizin in den Tropen, Heilen und Pflegen mit tropischen Pflanzen, Arzneimittel und Kosmetika selbst herstellen, self-published, Winnenden, 2012

54. Hoffman, U.; Weber, B.: Krank durch Amalgam – und was dann? Printed by: Grundblick; Druck & Verlag, Marburg-Moischt, 1996

55. Institute for construction biology and ecology (IBN – Institut für Baubiologie + Oekologie): Geldrollen im Blut durch Handystrahlung,

Jugend forscht – und findet wieder: Special issue of Wohnung + Gesundheit Nr. 115, Neubeuern, 2005

56. International doctors' appeal 2012, Mobilfunk gefährdet Gesundheit, Quelle: Freiburg Appeal 2012, umwelt-medizin-gesellschaft 25/4/ 2012, UMG Verlagsgesellschaft mbH Bremen, pp. 225

57. Ionescu, J. G.; Novotny, J.; Stejskal, V. et al.: Hohe Akkumulation von Übergangsmetallen im Brustkrebsgewebe, umwelt medizin gesellschaft, 4/i2006 MG – Verlagsgesellschaft, Bremen, pp. 269–273

58. Jennrich, P.: Schwermetalle, Ursache für Zivilisationskrankheiten, Edition CO'MED, Mediengruppe Oberfranken, Kulmbach, 2007

59. Jennrich, P.: Schwermetalle und Krebs, Auswirkungen bisher wenig beachtet, CO'MED trade journal 07/08, Mediengruppe Oberfranken, Kulmbach, 2008, pp. 6–11

60. Jennrich, P.: Europarat ruft auf, die Umweltbelastung durch Schwermetalle zu reduzieren, CO'MED trade journal 07/2011, Mediengruppe Oberfranken, Kulmbach, 2011, pp. 4–8

61. Karstädt, U.: entgiften statt vergiften, TAS, London, 2009

62. Kaufmann, G.: Offener Brief an Landrat Dietrich Kübler und Elsbeth Kniß, leitende Landwirtschaftsdirektorin, Fränkisch-Crumbach, 20.04. 2011

63. Klingel, R.: Therapeutische Apherese, Scientific Forum, Cologne Institute for Apheresis Research, 2010

64. Klinghardt, D.: Schwermetalle und ihre Wirkung auf die Gesundheit, Paracelsus conference at the ETH Zürich, 1998

65. Koos, B. J., Longo, L. D.: Mercury toxicity in the pregnant woman, fetus, and newborn infant. Am. J. Obstet Gynecol 126, Publisher: American Gynecological Society, Elsevier, 1976, pp. 390–409 [in English]

66. Kortenkamp, A.; Grahl, B.; Grimme, K. H.: Die Grenzenlosigkeit der Grenzwerte, zur Problematik eines politischen Instruments im Umweltschutz, Stiftung Ökologische Konzepte, Verlag C. F. Müller, Karlsruhe, 1990

67. Krahn-Zembol, W.: Tätigkeitsbericht mit dem Veröffentlichungsverzeichnis nach Themen und Stichwörtern geordnet, 36, Wendisch Evern, status on: 11/2011
68. Krude, H.; Reiners, C.: Schilddrüsenknoten bei Kindern und Jugendlichen, Monatszeitschrift Kinderheilkunde 156, Springer Medizin Verlag, Berlin, 2008, pp. 972–980
69. Kuusinen, A.: Der Gott stürzt seine Engel, Herausgeber: Wolfgang Leonhard Verlag Fritz Molden, Munich,1972
70. Langbein, K. et al.: Bittere Pillen, Nutzen und Risiken der Arzneimittel. Ein kritischer Ratgeber, Verlag Kiepenhauer & Witsch, Köln, 1983
71. Lechner, J.: Gesunde Zähne – gesunder Mensch, wie wichtig eine ganzheitliche Zahnheilkunde ist, Verlag Zabert Sandmann, München, 2009
72. Lewin, L.: Gifte und Vergiftungen: Lehrbuch der Toxikologie, Karl F. Haug Verlag, Ulm/Donau, 1962
73. Lipton, B.: Intelligente Zellen, Wie Erfahrungen unsere Gene steuern. Verlag KOHA GmbH Burgrain, 2012
74. Lown, B.: Die verlorene Kunst des Heilens, Anleitung zum Umdenken, Suhrkamp Taschenbuch Verlag, Stuttgart, 2004
75. Maciola, G.: Die Grundlagen der Chinesischen Medizin, Ein Lehrbuch für Akupunkteure und Arzneimitteltherapeuten, Verlag für chinesische Medizin Dr. Erich Wühr, Kötzting, 1994
76. Marschall, Ch.: Source: ZEIT Online, 19.09.2009, Verlag Der Tagesspiegel GmbH, Berlin, 2009
77. Martorell, I. et al: Human Exposure to Arsenic, Kadmium, Mercury and Lead from foods in Catalania, Biol. Trace Elem Res. United States, 2010 [in English]
78. Medizinisches Labor Bremen: Arbeits- und Umweltmedizinische Analysen, Bremen, 2007/2008, p. 134
79. Mebs, D.: Gifttiere: Ein Handbuch für Biologen, Toxikologen, Ärzte und Apotheker, Wissenschaftliche Verlagsgesellschaft mbH, Stuttgart, 2000

80. Merz, T.: Anerkennung von MCS in Gerichtsverfahren und der öffentlichen Meinung, umwelt-medizin-gesellschaft, 4/2007, UMG Verlagsgesellschaft, Bremen, pp. 1–7

81. Merzoug, K.; Gerhard, I.; Runnebaum, B.: Häufigkeiten und Voraussetzungen für therapieunabhängige Schwangerschaften bei Sterilitätspatientinnen. Geburtshilfe und Frauenheilkunde, 50, Georg Thieme Verlag Stuttgart, 1990, pp. 177–188

82. Mortimer, Ch. E.: Chemie: Das Basiswissen der Chemie, Verlag: Georg Thieme Verlag, Stuttgart, 1996, pp. 353–357

83. Mortimer, Ch. E.: Chemie: Das Basiswissen der Chemie, 6. edition, Georg Thieme Verlag 1996, Stuttgart, 1996, pp. 371–373

84. Mukherjee, S.: Der König aller Krankheiten Krebs – eine Biographie, DUMONT Buchverlag, Köln, 2012

85. Müller, K. E.: Die Umweltmedizin spielt in der Gesundheitspolitik der Parteien keine Rolle – auch in der von Bündnis 90/Die Grünen nicht, umwelt-medizin-gesellschaft, 24/ 2/2011, UMG Verlagsgesellschaft mbH Bremen, pp. 158–159

86. Müller, K. E.: Erschöpfung aus Sicht der Klinischen Umweltmedizin, umwelt-medizin-gesellschaft UMG 4/2012, UMG Verlagsgesellschaft mbH, Bremen, pp. 232–237

87. Müller, M.; Westphal, G.; Vesper, A. et al.: Inhibition of the human erythrocytic glutathione-S-transferase T1 (GSTT1) by Thiomersal, Int. J. Hyg. Environ Health 2003 (5–6): pp. 479–481 [in English]

88. Münkler, H.: Mitte und Maß, Der Kampf um die richtige Ordnung Rowohlt Taschenbuch Verlag, Hamburg, 2012

89. Mutter, J.: Gesund statt chronisch krank! Der ganzheitliche Weg: Vorbeugung und Heilung sind möglich, Fit fürs Leben, Verlag in der NaturaViva Verlags GmbH, Weil der Stadt, 2009

90. Natur & Heilen, monthly periodical, Gesund. Leben. Ganzheitlich, 4/2012, Munich, p. 3

91. Nogier, P.: Lehrbuch der Auriculotherapie, published by: Sainte – Ruffine, Maisonneuve, 1969

92. Nogier, P.: Wissenschaftliche Experimente des G.L.E.M., Zeitschrift für Akupunktur und Aurikulomedizin, 12 / 13, MDV-Verlag., Duderstadt 1976, pp. 45–46

93. NIS (Noven-Information-System) FAKT, Yusho Krankheit, from 21.10.1994,

94. Orphanet Deutschland: Handbuch Seltene Krankheiten, published by: Medizinische Hochschule Hannover, März 2007

95. Perger, F.: Unterschiedliche Entwicklungen der Schwermetallbelastungen (Pb, Cd, Hg) und ihre Therapie. Ärztezeitschrift für Naturheilverfahren Nr. 10, Elsevier/Urban & Fischer Verlag, München,1987, pp. 774–794

96. Petersen, E.: 20 Jahre Umwelt und Gesundheit aus Sicht einer Nichtregierungsorganisation, umwelt-medizin-gesellschaft 2/2010, UMG – Verlagsgesellschaft Bremen, p. 87

97. Philippi, V.: Bioenergetische Meditation nach Viktor Philippi, Broschüre, Herausgeber: Forschungs- und Lehrakademie für Bioenergetik und Bioinformatik Viktor Philippi, Sohland an der Spree, 2009

98. Prawda, W.: Endlager Mensch – auch Sie sind vergiftet. Ursachen, Diagnose, Therapie, Verlag Books on demand GmbH, Norderstedt, 2011

99. Rath, M.; Niedzwiecki, A.: Krebs – Das Ende einer Volkskrankheit, volume 1, Der wissenschaftliche Durchbruch, Dr. Rath Health Foundation, Vertrieb, Dr. Rath Education Services B.V., NL –Heerlen, 2011

100. Resident Health Management Survey of Fukushima Prefecture, Report on thyroid cysts and nodules in children, from 26.4.2012

101. Ruprecht, J.: Dimaval R (DMPS) DMPS – Heyl, R. scientific product monography, Heyl chem.- pharrm. Fabrik GmbH & Co KG, Berlin, January 1997

102. Scheiner, H. C.: Der 100 Milliarden Euro Skandal. Gesundheit für Alle! Die große Reform durch Naturmedizin (2003): Leidet der Patient nicht schon genug? Matrix 3000 Zeitschrift, Michaels Verlag, Peiting, 2003

103. Schiwara, H.W. et al: Bestimmung von Kupfer, Quecksilber, Methylquecksilber, Zinn, Methylzinn und Silber in Körpermaterial von Amalgamträgern, Klinisches Labor Bremen, 9/92

104. Schöndorf, E.: Amalgamurteil gegen Degussa, Staatsanwaltschaft bei dem Landgericht Frankfurt am Main, 31.05.1996

105. Schöndorf, E.: Open letter to the federal government and the members of the German parliament, Bad Vibel, 10.06.1999

106. Schuhardt, E.; Kopelex, L.: Die Stimmen der Kinder von Tschernobyl: Geschichte einer stillen Revolution, Herder Verlag, Feiburg, 1996

107. Schulte-Uebbing, C,: Angewandte Umweltmedizin, Sonntag Verlag Stuttgart, 1996, p. 132

108. Soddemann, W.: Mikrobiologische und chemische Trinkwasserbelastung, lecture at the 11. conference on environmental medicine in Hamburg, 19.–21.10.2012

109. Steinert, J.: Das Gift wabert aus den Wänden: Zeitschrift für Umweltmedizin Nr. 41, 2001, p. 230

110. Stock, A.: Die Wirkung von Quecksilberdampf auf die oberen Luftwege; Naturwissenschaften 23, 1935, pp. 453–356

111. Stock, A.: Die Gefährlichkeit des Quecksilberdampfes und der Amalgame, Medizinische Klinik, Wochenschrift für Praktische Ärzte, 22, 1926, pp. 1209–1212 and 1250–1252

112. Stock, A.: Die chronische Quecksilber- und Amalgamvergiftung; Archiv für Gewebepathologie und Gewebehygiene 7, 1936, p. 388–413

113. Straube, R.: Innovative Strategien in der Behandlung von MCS-CFS-SBS, I Zeolithbiotechnologie und Doppelmembranfiltrationsapherese, Umwelt-Echo issue 25, IV. quarter, 2005

114. Straube, R., Donate, H. P.: Die Doppelmembranfiltrationsapherese als Behandlungsoption bei Erkrankungen aus der Umweltmedizin, umwelt-medizin-gesellschaft, 23/1/2010 UMG – Verlagsgesellschaft, Bremen, pp. 9–14

115. Teufel, M. et al.: Chlorinated Hydrocarbons in Fat Tissue: Analyses of Residues in Healthy Children, Tumor Patients, and Malformed Children; Arch. environ Contam. Toxicol 19, 1990, pp. 646–652

116. Teufel, M.; Böhm, J.; Niesen, K.: Belastung unserer Kinder mit Pestiziden, PCB und potentiell kritischen Anionen, Monatszeitschrift Kinderheilkunde 139, Springer Verlag, Berlin 1991, pp. 442–449

117. Teufel, M.: Bedeutung der Umweltbelastung mit polychlorierten Biphenylen (PCB) im Kindesalter, Der Kinderarzt Nr. 12, Springer Verlag Berlin, 1997

118. Thiede, W.: Mythos Mobilfunk, Kritik der strahlenden Vernunft, Verlag oekom München, 2012

119. Van Praagh, J.: Meditation für innere Kraft und Lebensfreude, Wilhelm Heyne Verlag, München, 2007

120. Wassermann, O.; Weltz, M.: Kiel expert witness report, reference number: Js 17084.4/91 Christian-Albrecht University in Kiel, Institute for Toxicology,1995

121. Westphal, G.; Schmuch, A.; Schulz, T. et al.: Homozygous gene deletions of the glutathione-S-transferase M1 and T1 are associated with thiomersal sensitization, Arch. Occup Environ Health 73, 2000, pp. 384–388 [in English]

122. Westerhoff, B. et al.: A comparative electrochemicla in vito evaluation of the corrosion behaviour of dental amalgams, J. Oral Rehabil. 1995, pp. 121–127 [in English]

123. White, F. M. et al.: Chemicals, birth defects and stillbirths in New Brunswick: associations with agriculutry activity. Can. Med. Assoc. J. 138, 1988, pp. 117–124 [in English]

124. Wiebe, J. P.; Barr, K. J.; et al.: Effect of prenatal and neonatal exposure to lead on gonadotropin receptors and steroidogenesis in rat ovaries. J.Toxicol. Environ Health, 24, 1988, pp. 461–476 [in English]

125. Wittke, J.W.: Arbeits- und Umweltmedizinische Analysen, Bremen medical laboratory, 2007/2008

126. Wortberg, W.: Trennung vermehrungsanregender Stoffe aus Hefeextrakten von Saccharomyces cerevisiae, Pharmakologischen Institut in Marburg, Dissertation for the attainment of the degree of Medical Doctor at the Marburg Philipps University, 1968

127. Wortberg, W.: Ursache und Verhinderung von Oberschenkelhalsbrüchen bei älteren Menschen. Die Vorstellung eines Stoßneutralisators zur Verhinderung von Schenkelhalsfrakturen bei älteren Menschen, German society for gerontology and geriatrics, Berlin, 1986

128. Wortberg, W.: Hüft-Fraktur-Bandage zur Verhinderung von Oberschenkelhalsbrüchen bei älteren Menschen. Der Oberschenekelhalsbruch, ein biomechanisches Problem, Zeitschrift für Gerontologie, Springer Verlag, 1988, pp. 169–173

129. Wortberg, W.: In-vivo-Untersuchungen mit einer Senioren-Sicherheitshose zur Verhinderung von Oberschenkelhalsbrüchen bei älteren Menschen, Geriatrische Forschung Vol. 8. Nr. 1, MMV Verlag, Munich, 1998

130. Wortberg, W.: Intrauterine Fruchtschädigung durch Schwermetallbelastung der Mutter, umwelt-medizin-gesellschaft / 19/4/2006, UMG – Verlagsgesellschaft Bremen pp. 274–280

131. Wortberg, W.: Environmental Medicine and Health in Industrial Nations: Cases from Germany, in: EDMONDSON R, RAU H (eds): Environmental Argument and Cultural Difference, Peter Lang AG, Bern, Berlin, Frankfurt am Main, New York, Vienna, 2008, pp. 233–258

132. Wortberg, W.: Einfluss von Schwermetallen, genetischen und immunologischen Faktoren auf die Entstehung von bösartigen Tumoren. Teil 1: Einfluss von Schwermetallen, CO'MED 01/2012, Mediengruppe Oberfranken, Kulmbach, pp. 58–60 and 69–71

133. Wortberg, W.: Einfluss von Schwermetallen, genetischen und immunologischen Faktoren auf die Entstehung von bösartigen Tumoren. Part 2: Einfluss von immunologischen und genetischen Faktoren. Neue Diagnostik- und Therapiemöglichkeiten, CO'MED 03/2012, Mediengruppe Oberfranken, Kulmbach, pp. 46–49

134. Wortberg, W.: Metalle, die verborgenen Ursachen von bösartigen Tumoren und chronischen Erkrankungen, umwelt-medizin-gesellschaft, UMG – Verlagsgesellschaft mbH, Bremen, 1/2013, pp. 39–44

135. Wortberg, W.: Nach 12 Jahren: Anerkennung einer toxischen Hirnschädigung durch Umweltgifte, umwelt-medizin-gesellschaft 22/2, UMG – Verlagsgesellschaft Bremen, 2009, pp. 139–147

136. Ziff, S.; et al.: Amalgam Die toxische Zeitbombe Zahnmedizin im Umbruch. Was wissen wir über Amalgam? Gibt es sichere Alternative? Copyright by Felicitas, Hübner Verlag, Waldeck, 1985

www.ingramcontent.com/pod-product-compliance
Ingram Content Group UK Ltd.
Pitfield, Milton Keynes, MK11 3LW, UK
UKHW021822140426
5217IPUK00004B/51